AF894236

The Evolutionists

The Evolutionists

Aristotle to Zoroaster via Darwin

Peter Sheldon

BREWIN BOOKS

BREWIN BOOKS
19 Enfield Ind. Estate
Redditch
Worcestershire
B97 6BY
www.brewinbooks.com

Published by Brewin Books 2025

© Peter Sheldon, 2025

The author has asserted his rights in accordance with the Copyright, Designs and Patents Act 1988 to be identified as the author of this work.

All rights reserved. No part of this publication may be reproduced, stored in a retrieval system, or transmitted in any form or by any means, electronic, mechanical, photocopying, recording or otherwise, without the prior permission in writing of the publisher and the copyright owners, or as expressly permitted by law, or under terms agreed with the appropriate reprographics rights organisation. Enquiries concerning reproduction outside the terms stated here should be sent to the publishers at the UK address printed on this page.

The publisher makes no representation, express or implied, with regard to the accuracy of the information contained in this book and cannot accept any legal responsibility for any errors or omissions that may be made.

A CIP catalogue record for this book is available from the British Library.

ISBN: 978-1-85858-772-1

Printed and bound in Great Britain.

Contents

Ignorance more frequently begets confidence
than does knowledge – Charles Darwin –
The Descent of Man

Give Chance a chance that it may
chance upon the right chance – Anon

A race is a grouping of humans based on shared
physical or social qualities into categories
generally viewed as distinct by society.

On the origin of species, or the preservation of
favoured races in the struggle for life
– Charles Darwin, 1859

Preface

Evolution is defined in the Oxford Dictionary of English (2005) as:

> The process by which different kinds of living organism are believed to have developed from earlier forms during the history of the earth.

There were some thinkers in Ancient Greece who had proposed organic evolution, but the idea was firmly rejected by those who held religious views. Lamarck felt that the environment drove transformation in a fairly swift timescale. This however did not appear to be the case after Lyell deduced from fossils that the timescale was vastly longer than anything envisaged by religion. Acceptance of the possibility of extreme duration of time led Erasmus (Charles Darwin's grandfather) and Charles to propose what is now referred to as Darwin's theory of evolution. This was simultaneously espoused by **Alfred Wallace** (1823-1913). The role of inheritance (**Gregor Mendel** 1822-1884), and genes (DNA) has led to the fascinating concept of evolution now becoming revealed at a molecular level, though tantalisingly still hiding several secrets.

This book starts with the Greeks then covers the lives of three generations of the Darwins, and beyond, over four centuries. Wallace is accorded his rightful place in this scheme. Further chapters examine Darwinian evolutionary theory and re-examine the Lamarckian concept, in the light of 21st-century ideas.

The Darwins, who justifiably have occupied the first place in most people's minds are described. Firstly **Erasmus**, (1731-1802), grandfather of

Charles, and who lived initially in Lichfield. He was a man of many talents (including the pursuit of the opposite sex), who invented an artificial voice box, a means of steering for horsedrawn coaches, serious poetry, as well as a theory that all life originated from primitive aquatic forms. His carriage bore the motto, *e conchis omnia* (from snails, everything).

In addition, he was a much respected doctor. In the next generation, we meet his son **Robert Waring** (1768-1848), best known as a successful physician in Shrewsbury who though never to achieve the success attributed to his father, or his son, nevertheless somehow is credited with having presented a paper to the Royal Society on the effects of ocular spectra. This acquired him fellowship of this society, as a result of which he could add the letters FRS to his name, already adorned with MD. His inclusion in this book is tenuous, and purely due to his family linkage as the son of Erasmus, and the father of Charles. His contribution, if any, to evolution is unknown. Finally, his son **Charles**, (1809-1882), born in Shrewsbury, educated (though not happily), at Shrewsbury School. Of the three Darwin generations, the most famous and highly respected undoubtedly was Charles, yet he was the only one who was not medically qualified.

The 1798 treatise *An Essay on the Principle of Population*, by English political economist Thomas Robert **Malthus**, inspired Charles Darwin and Alfred Russel Wallace to work out the mechanics of natural selection based on Malthus' observation that populations tend to increase geometrically (2, 4, 8, 16 ...), whereas food reserves grow arithmetically (2, 3, 4, 5 ...), leading to competition for scarce resources and differential reproductive success, the driver of evolution.

Sir Charles Lyell, 1st Baronet, FRS (1797-1875) was a Scottish geologist who demonstrated the power of known natural causes in explaining Earth's history. He was a close friend of Charles Darwin, and contributed significantly to Darwin's thinking on the processes involved in evolution. Lyell helped to arrange the simultaneous publication in 1858 of papers by Darwin and Alfred Russel Wallace on natural selection, despite his personal religious qualms about the theory. He later published evidence from geology of the time Man had existed on Earth. He is best known as the author of *Principles of Geology* (1830-1833), which presented to a wide public audience the idea that Earth was shaped by the same natural processes still in operation today, operating at similar intensities. The philosopher William Whewell termed this gradualistic view *uniformitarianism* and contrasted it with *catastrophism*, which had been championed by Cuvier and was better accepted in Europe.

The driving force was, and still is, Man's enquiring mind. First how, and then why. The answer to the first e.g. the concept of the evolution of species, leads on to the thorny question of why. At this point the clergy picks up its ears, and the seeds of conflict are sown. The various religions had had it all their own way, until Charles Darwin subtly implied an alternative explanation of the how. This naturally led to the why, and the germination of *Horticus conflictus.* The name Darwin is undoubtedly the best known, but other thinkers especially **Alfred Russel Wallace** also made significant contributions, he is arguably the equal of Darwin. The subsequent way forward relied on science, of which undoubtedly the work of **Gregor Mendel**, the Austrian monk, pointed the way to the existence of genes and inheritance.

2013 marked the centennial of the death of Wallace. In 1857 he wrote to Charles Darwin, pre-empting his idea of natural selection, and Darwin, a true Victorian gent, suggested that they present a joint paper on the idea. Wallace, in turn, became his devotee, and is remembered mainly in that role.

In truth, Wallace did far more. He started off life as a surveyor, and set out the street plan of Llandrindod Wells, in the Welsh borders. He then designed a council building in Neath. He travelled in South America, where his brother died and the ship carrying his specimens caught fire and sank.

In the midst of all this, Wallace founded the **science** of biogeography, the attempt to make sense of the distribution of animals and plants. In a map made in 1876 he defined six great regions of life; and in so doing predicted, without realising it, the greatest discovery of geology in the past century, the idea of continental drift.

In Indonesia, he discovered a discontinuity in the patterns of animal distribution. A line – later called the **Wallace Line** – drawn across the archipelago, with the Philippines, Borneo, Java and Bali to its north and west; and Lombok, Sulawesi and New Guinea to the south and east – separated two biological worlds. One had ties to Asia, with its fruit-thrushes, weaver birds and tigers, and the other to Australia, with cockatoos, honeyeaters and kangaroos. In some places, just 15 miles of sea separated the two provinces.

Why was this? In Wallace's day the oceans were full of hypothetical land bridges, now sunk, which were said to have linked, for example, Australia, Africa and South America, to explain how flightless birds had got to all three. Wallace, too, appealed to the imagined landscape of Lemuria, a continent sunk in the Pacific, to explain the connection between Indonesia and Madagascar.

Shortly before Charles Darwin, there was an alternative theory with regard to inheritance, namely that of **Lamarck**. Whereas Darwinian evolution relied on natural selection alone, if the inherited formula (whatever that consisted of, and he didn't know), worked, then the species could survive, reproduce, and nurture unto the next generation. If it didn't, it wouldn't. There was no after-market modification (as in today's automobile industry). Lamarck by contrast implied that it was all due to after-market influence upon a basic inherited formula. Unfortunately the Lamarckian concept of the giraffe's long neck being inherited as a consequence of the parent's stretching up to reach the tasty leaves, resulted in ridicule reinforced, for a while, by the exclusive acceptance of the DNA to RNA to protein paradigm, to the exclusion of all else. This paradigm is known as the **Weismann line**. Fascinatingly, all this is nowadays challenged, not least by the field of **epigenetics**.

This work will lead the reader through many centuries and three generations of Darwins, two of whom were doctors and one, arguably the greatest, a biologist. Hopefully it will be an interesting and intelligible read. References are few and far between, and their absence permits the prose to flow. The author confesses to the use of cutting and pasting various tracts from other sources which are deemed reliable, and whose style would be difficult to alter without justification.

Chapter 1

Pre-Darwin – Aristotle to Lucretius

Aristotle (385-323 BC)

Aristotle

Aristotle was born in the Macedonian region of north-eastern Greece in the small city of Stagira (hence the nickname 'the Stagirite'). At about the age of seventeen Aristotle was sent to Athens to study in Plato's Academy, then a pre-eminent place of learning in the Greek world. He is regarded as one of the greatest philosophers of all time.

Aristotle's approach to the study of logic was not as an end in itself, but rather a view to its role in human inquiry and explanation. He regarded logic as a tool which he thought made an important but incomplete contribution to science and the truth of opinion. Science he felt needed arguments, not just deductions.

He queried the role of male semen in the developing embryo, where did it go, did it contribute some inner force, and what was its role in the soul? Did it provide nutriment to this end?

> Thus while it is necessary for the female to provide a body and a material mass, it is not necessary for the male, because it is not

within the work of art or the embryo that the tools or the maker must exist. ***While the body is from the female, it is the soul that is from the male.***

(Author's comment: How would today's women react to this assumption?!)

And yet the question may be raised why it is that, if indeed the female possesses the same soul and if it is the secretion of the female which is the material of the embryo, she needs the male besides instead of generating entirely from herself. The reason is that the animal differs from the plant by having sense-perception; if the sensitive soul is not present, either actually or potentially, and either with or without qualification, it is impossible for face, hand, flesh, or any other part to exist; it will be no better than a corpse or part of a corpse. If then, when the sexes are separated, it is the male that has the power of making the sensitive soul, it is impossible for the female to generate an animal from itself alone, for the process in question was seen to involve the male quality.

Aristotle believed that species were fixed by divine design. As such, he did not believe in evolution. However, there were earlier Greek philosophers who did believe in the ability for species to change and the creatures that did exist represented some form of intermingling and mixing of elements of animals.

Aristotle's *De Generatione Animalium* (on the Generation of Animals), was produced during the latter part of the 4th century BC. The following is taken from the commentary of Cera R. Lawrence, *Embryo Project Encyclopedia* (2010-10-02). ISSN: 1940-5030:

This book is the second recorded work on embryology as a subject of philosophy, being preceded by contributions in the Hippocratic corpus by about a century. It was, however, the first work to provide a comprehensive theory of how generation works and an exhaustive explanation of how reproduction works in a variety of different animals. As such, ***De Generatione*** was the first scientific work on embryology. Its influence on embryologists, naturalists, and philosophers in later years was profound. Among these were ***Hieronymus Fabricius, William Harvey, St. Thomas Aquinas****, and* ***Charles Darwin****.*

The Aristotelian approach to philosophy was teleological, and involved analysing the purpose of things, or the cause for their existence. These causes were split into four different types: final cause, formal cause, material cause, and efficient cause. The final cause was what a thing existed for, or its ultimate purpose. The formal cause was the definition of a thing's essence or existence, and Aristotle stated that in generation, the formal cause and the final cause were similar to each other, and could be thought of as the goal of creating a new individual of the species. The material cause was the stuff a thing was made of, which in Aristotle's theory was the female menstrual blood. The efficient cause was the "mover" or what caused the thing's existence, and for reproduction Aristotle designated the male semen as the efficient cause. Thus, while the mother's body contained all the material necessary for creating her offspring, she required the father's semen to start and guide the process.

De Generatione consists of five books, each containing multiple chapters. Books I and II are of most interest to embryology. Book III is a comparative study of zoology that applies principles from Book II to different species of animals. Book IV contains miscellaneous information about aspects of reproduction, such as how heredity works and birth defects occur. Book V compares the characteristics that all animals share, and is primarily a discussion of sensory organs and the physical appearance of animals, focusing on characteristics like hair, coloration, voice, and teeth.

Book I begins with a discussion of maleness and femaleness, which Aristotle separates into active and passive principles, again underlining male semen as the efficient cause of generation and the female menstrual blood as the material cause. He gives reasons for the various shapes of male and female reproductive organs. Chapters 8 through 11 are a comparative anatomy of uteri and eggs, in which Aristotle classifies animals by uterine form and position and as well as whether the species is viviparous (live birth), oviparous with hard-shelled eggs, oviparous with shell-less eggs, or ovoviviparous as in fishes that hatch from eggs contained within the mother. Chapters 12 and 13 discuss why uteri are internal, why testes are not always internal, and how the urinary anatomy is related to the genitals. In chapters 14 through 16 Aristotle describes copulation in some mobile non-sanguinous animals (insects, crustaceans, and cephalopods) that do not have red blood. He returns to the definition of seminal fluids in chapters 17 through 20. He refutes the idea that the male seed is composed of particles from every part of the parent's body – a theory that was resurrected in the eighteenth century as "pangenesis" as support for the concept of generation known as spermist preformationism, and later used by Darwin to account for heredity. Aristotle asserts that the female does not have a seminal fluid, but rather that her contribution to her offspring is only in the form of the nutritive menstrual blood that is the material of generation.

Book II begins with a systematic classification of all animals based on embryological traits. An informative chart of this system can be found in Needham's History of Embryology in the section about Aristotle. This first chapter also includes a discussion of epigenesis versus preformationism, though Aristotle does not use those terms. He refutes the idea that the embryonic parts could exist in the semen (preformationism) since in his observations of the embryos of many different animals, he had clearly seen the heart formed first and other parts formed only gradually afterwards (epigenesis). Aristotle's theory of epigenesis remained the dominant model for embryogenesis until the middle of the seventeenth century.

Chapter 3 of Book II defines the degree of "aliveness" at various stages of embryological development. This is the section in which Aristotle discusses three different types of human souls: a nutritive soul, imbued from the very beginning; a sensitive soul, imbued later; and finally the intellective soul, imbued forty days after conception for a male embryo and eighty days for a female embryo. The nutritive soul, also called the vegetative soul, is the essence possessed by all living things, including plants, and can be considered the lowest level of soul. The sensitive soul is what separates plants from animals, and provides animals the ability to move and to interact with the world around them. The intellective soul is what separates humans from all other animals, and allows humans to think and reason.

Chapters 6 and 7 of Book II explain the process and order of embryo development. In Aristotle's account, based in part on observations of chicken eggs, the heart is formed first, the rest of the internal parts next, and finally the external parts. Anterior parts are formed before posterior parts. Aristotle asserts that the order of formation is determined by the importance of each part to the adult form, making it clear that the process of embryo development is aimed towards the final cause of producing an individual of its species. He states that no part of the forming embryo creates any other part, which denies other philosophers' assertions that the heart causes the formation of the rest of the body. The emphasis Aristotle places on the orderly development of the parts of the embryo is a precursor to the processes proposed by two later scientists. The laws of embryology proposed by Karl Ernst von Baer and the recapitulation theory of Ernst Haeckel both placed a similar emphasis on the epigenetic development of the embryo.

Book III discusses in detail the reproductive process of each class of animal other than the viviparous group identified by Aristotle in chapter 1 of Book II. It begins with a discussion of avian eggs, which, like reptilian eggs, are "perfect eggs" not requiring a second fertilization after they are laid. He explains how so-called unfertilized "wind eggs" are produced by birds such as chickens and asserts that the yolk of the egg is the nutritive substance and not the white albumen. Chapter 3 is a similar discussion of oviparous fishes, which lay imperfect eggs that must be fertilized by the male a second time. Aristotle maintains that all animals with two sexes must engage in sexual union and fish are no exception, though according to Aristotle they copulate very quickly. Chapter 8 discusses cephalopod procreation, asserting again that sexual union has been observed in squid. Chapter 9 is about insects, which Aristotle says sometimes arise spontaneously and sometimes generate offspring. Generation of offspring is done by producing a scolex, which he says is an egg laid too soon, as evidenced by the fact that it grows in size before hatching. He attributes metamorphosis to the natural outcome of spontaneous generation. Those metamorphic insects that arise spontaneously, in Aristotle's observations, rather than from beings that look like themselves, cannot produce offspring that look like themselves. Their offspring must be perfected through metamorphic stages. Chapter 10 is about bees, which Aristotle says are very difficult to explain. From his observations, he concludes that queen bees produce both workers and other queens, workers produce drones, and drones produce nothing. Book III ends with a chapter about testacea (bivalve and univalve molluscs), which Aristotle says are like plants because they do not move during the entirety of their life.

Book IV contains an explanation of why offspring resemble their parents. In Aristotle's view, heredity is the result of the form carried by the paternal seed. If the development of the embryo proceeds perfectly successfully, the offspring will be male, and closely resemble his father. Female offspring are the result of less perfect development and resemble their mothers. Males who resemble their mothers or females who resemble their fathers are examples of development in between these first two levels, and even less perfect developments will cause the offspring to resemble more and more distant ancestors. Aristotle's teratology describes the defects that occur with extremely disturbed gestations, a model of birth defects that persisted for centuries after Aristotle. This view of heredity and birth defects, along with the writings by Galen on the same subject, persisted through the Middle Ages in both Europe and the Islamic Caliphate. The reproductive role that Aristotle ascribed to the female, who contributes only the material for the offspring and influences the form only when gestation is "less perfect", influenced reproductive medicine for nearly two millennia.

The remainder of De Generatione is not of great concern to embryology. Book IV concludes with miscellaneous information about milk production, how natural births proceed, and the reasons for differing periods of gestations in different animal species. Book V consists mostly of a comparison of the sense organs in animal species, as well as traits in hair coloration, voice pitch, and teeth.

Aristotle's embryological work was highly influential after his death and throughout the Middle Ages. Two notable medieval scholars influenced by De Generatione were St. Albertus Magnus and St. Thomas Aquinas. The latter was specifically influenced by Aristotle's ideas of ensoulment, and developed a theory of delayed hominization, in which first the vegetative soul enters the foetus, then the animal soul, then the human soul at around forty days for males and eighty days for females. Aquinas and St. Augustine before him both used this idea to define when human life begins, and their ideas were used by the Catholic Church to promulgate anti-abortion law.

The European Enlightenment era, so concerned with experimentalism, saw a revival in interest in the observational aspects of Aristotle's work. ***William Harvey***, anatomist and author of his own Exercitationes de Generatione Animalium (1651), was an adherent of the Aristotelian idea of epigenesis, influenced by his teacher Hieronymus Fabricius. Both Fabricius and Harvey built on Aristotle's work with their own observations, making corrections to the original where necessary. In his 1672 De Mulierum Organis Generationi Inservientibus, Reinier de Graaf, discoverer of the ovarian Graafian follicles and champion of the female role in reproduction, wrote about the clear evidence that offspring often resembled both parents and discussed the problems thus created for Aristotle's theories. In the latter part of the 19th century, Charles Darwin famously wrote that Aristotle had been one of his chief influences.

Aristotle's De Generatione Animalium marked the beginning of the history of Western embryology. Though a great deal of Aristotle's work in De Generatione was based on observation, these observations were not made in the controlled style expected of scientists in modern times, and many of the conclusions he drew were based on anecdotes and second-hand information. The modern reader is privy to much more information about the lives and habits of members of the animal kingdom than Aristotle was, and many of his assertions thus seem absurd. However, Aristotle's concern with observation, and his method of examining as many different kinds of animals as possible and comparing and contrasting their traits, was a precursor to the modern style of investigation. Before Aristotle, ideas about generation and reproduction were confused, more than half supernatural, and by no means comprehensive. In writing De Generatione, Aristotle left a number of unsolved problems of reproduction for researchers who followed him to build on, while providing a framework in which to conduct that research. Before De Generatione, those who wanted to study reproduction had to start from scratch; after De Generatione, future research might fit into Aristotle's scheme and thus become part of the body of scientific knowledge.

There is no doubt that book IV will cause the hackles to rise.

Hippocrates (460-370 BC)

Hippocrates of Kos also known as Hippocrates II, was a Greek physician of the Age of Pericles (Classical Greece), who is considered one of the most outstanding figures in the history of medicine. He is often referred to as the "Father of Medicine" in recognition of his lasting contributions to the field as the founder of the Hippocratic School of Medicine. This intellectual school revolutionized Ancient Greek medicine, establishing it as a discipline distinct from other fields with which it had traditionally been associated thus establishing medicine as a profession.

Hippocrates

However, the achievements of the writers of the Corpus, the practitioners of Hippocratic medicine, and the actions of Hippocrates himself were often conflated; thus very little is known about what Hippocrates actually thought, wrote, and did. Hippocrates is commonly portrayed as the paragon of the ancient physician and credited with coining the Hippocratic Oath, which is still relevant and in use today. He is also credited with greatly advancing the systematic study of clinical medicine, summing up the medical knowledge of previous schools, and prescribing practices for physicians through the Hippocratic Corpus and other works.

Hippocrates is credited with being the first person to believe that diseases were caused naturally, not because of superstition and gods. Hippocrates was credited by the disciples of Pythagoras of allying philosophy and medicine. He separated the discipline of medicine from religion, believing and arguing that disease was not a punishment inflicted by the gods but rather the product of environmental factors, diet, and living habits. Indeed there is not a single mention of a mystical illness in the entirety of the Hippocratic Corpus. However, Hippocrates did work with many convictions that were based on what is now known to be incorrect anatomy and physiology, such as Humorism.

Hippocrates' contributions revolutionized the practice of medicine; but after his death the advancement stalled. So revered was Hippocrates that his teachings were largely taken as too great to be improved upon and no significant advancements of his methods were made for a long time.

So what is the relevance of his inclusion in this book?

It is based on the following statement by Charles Darwin in a letter replying to one written by a Dr William Ogle.

In 1868, Charles Darwin, already famous for his radical theory on evolution, made a surprising admission, acknowledging the similarities between his theories and those of Hippocrates, the famous Greek physician of the fifth century BC. In reply to a letter, now unfortunately lost, sent by Dr William Ogle (Superintendent of Statistics to the Registrar-General) Darwin declared the following:

> ...I wish I had known of these views of Hippocrates before I had published, for they seem almost identical with mine - merely a change of terms - and an application of them to classes of facts necessarily unknown to the old philosopher. The whole case is a good illustration of how rarely anything is new.
>
> ...Hippocrates has taken the wind out of my sails, but I care very little about being forestalled. I advance the views merely as a provisional hypothesis, but with the secret expectation that sooner or later some such view will have to be admitted.
>
> ...I do not expect the reviewers will be so learned as you otherwise, no doubt, I shall be accused of wilfully stealing Pangenesis from Hippocrates, for this is the spirit some reviewers delight to show.

Unfortunately one cannot verify this amazing assertion, but judging from the many other favourable attributes paid to Hippocrates, it is very likely to be true.

Empédocles de Agrigento (ca. 490-430 BC)

Empédocles

Empédocles was a Greek pre-Socratic philosopher and a native citizen of Akragas, nowadays Agricento, then a Greek city in Sicily. Empédocles' philosophy is best known for originating the cosmogonic theory of the four classical elements. Cosmogony is any model concerning the origin of either the cosmos or the universe. He also proposed forces he called Love and Strife which would mix and separate the elements, respectively.

Temple at Agricento, Sicily

The Empédocles papyrus, held since 1904 at the University of Strasbourg

Influenced by Pythagoras (died c. 495 BC) and the Pythagoreans, Empédocles challenged the practice of animal sacrifice and killing animals for food. He developed a distinctive doctrine of reincarnation. He is generally considered the last Greek philosopher to have recorded his ideas in verse. Some of his work survives, more than is the case for any other pre-Socratic philosopher. Empédocles' death was mythologized by ancient writers, and has been the subject of a number of literary treatments.

In his philosophical poem *Nature,* he reduced the whole variety of things to four elements of nature: earth, water, air and fire. This doctrine remained for many centuries in ancient and medieval philosophy.

Empédocles explained the unification and division of the elements by the action of two forces in dispute: attraction and revulsion. The predomination of one or the other force explains the different stages of development of the Universe. His conjecture of the logical evolution of living beings *by virtue of the natural selection of more viable combinations* was of great historical importance, in a sense pre-dating Charles Darwin.

Alain Martin, Faculty member, papyrologist, of Brussels University has claimed that a piece of papyrus, held at the University of Strasbourg since 1904, contains 70 lines of text by Empédocles. If true, it will be a major find, as very few manuscripts by pre-Socratic philosophers survive.

The following is by **Richard Janko**. University of Michigan, Ann Arbor:

> The papyrus is of Egyptian origin and has been dated to the 1st century BC. Although Empédocles is cited and quoted by other writers, including Aristotle, only two poems survive from his original writings, making the discovery especially important.

Empédocles is credited with having founded modern medicine as well as prefiguring Darwin, in a poem which talks of survival of the fittest among competing amalgams of limbs and body parts. His greatest fame, however, is in the legend that he leapt into the volcanic crater of Mount Etna to prove his divinity. Unfortunately he did not survive to tell the tale.

The Strasbourg papyrus contains over 70 complete or partial lines of Empédocles' poem. Its editors faced an extraordinarily difficult task in reconstructing it; they succeeded admirably in joining 47 of the original 52 fragments into six larger pieces, leaving only five small scraps.

The main thesis of the papyrus is (numbers refer to the line of the papyrus):

(1): there is a cycle where one comes from many, and many from one, because of Love and Strife (233–44). EXHORTATION: *listen to me, as learning improves the mind (245). REPRISE: there is a cycle involving the four elements, Love and Strife (246–51).* EXHORTATION: *use your imagination and do not sit dumbstruck (252).* PROOF: *Love exists, and is called 'Joy' and 'Aphrodite' by mortals (253–6).* EXHORTATION: *pay attention to my truthful words (256).* MAIN THESIS (2): *the elements predominate by turns over time (257–60). They do not come to be or perish (261–4). They constantly change but are constantly alike (265–6).* MAIN THESIS (3): *We come together in Love, but separate in Strife (267–8). This includes everything: plants, human beings, beasts, fish and gods (269–72). Everything constantly changes over the eons (273–7).*[50] *– Example: everything constantly changes, sun, moon and stars, in a cycle (278–82).*[51] *But when Strife reaches the middle of the vortex and Love arises there, then everything comes together there as one (288–90).* EXHORTATION: *Try to understand from what you can see (291–2).* PROMISE (1): *I shall show how Love makes things grow, come together and develop, and how things still remain from this creation (293–5). – Examples: beasts, humans, and plants. (296–8)* EXHORTATION: *From them try to understand. (299)* PROMISE (2): *You will see that things that come together and develop contrive divisive deeds because of Strife (300–1).*[52] *– First example: human beings. In youth our limbs come together in Love as one, but at life's end they wander, sundered by evil Strife (302–5). – Second example: plants, fish, beasts and birds (306–7).*

A double tale I'll tell. At one time one thing grew to be just one
from many, at another many grew from one to be apart.
Double the birth of mortal things, and double their demise.
Union of all begets as well as kills the first;
the second nurtures them but shatters as they grow apart.
And never do they cease from change continual,
at one time all uniting into one from Love,
while at another each is torn apart by hate-filled Strife.
In the way that many arise as the one again dissolves,
in that respect they come to be and have no life eternal;
but in the way that never do they cease from change continual,
in this respect they live forever in a stable cycle.

But hear my words; to learn augments the mind.
For as I said when I set forth my story's aims,
a double tale I'll tell. At one time one thing grew to be just one
from many, at another many grew from one to be apart,
fire, water, earth and the unreached height of air,
and cursèd Strife apart from them, their match in every way,
and Love among them, equal in her size and in her breadth.
With mind regard Her, and sit not with eyes bedazed.
Even mortals hold that She's implanted in their joints;
through Her they think of love and do conjoining deeds,
naming Her 'Delight' and 'Aphrodite' too.
No mortal man has learned that She revolves
among these things; but hear from me this truthful tale.
For all these things are equal and alike in age,
but each rules separate domains, each has its haunts

and lords it in its turn as time rolls round.
Beyond these nothing comes to be or perishes.
For if they died continually, no longer would they be.
What could increase this whole, and from what source?
How too could it be destroyed, since nothing lacks in these?
But these are what there is, and running through each other
they suffer change continual but always are alike.
In Love we come together in one world;
in Hatred many grew from one to be apart,
whence all that was, and is, and shall at some time be
blossomed as trees, as men, as women too,
as beasts, as birds, as fish that water rears,
as well as gods who ages live and greatest honours have.
In Her they never cease to swirl in constant flux
with frequent whirlings . . . .
relentlessly, and never do they cease . . . .
But many ages previous must elapse . . . .

before their motions alter . . . .
They never cease in any way to swirl in constant flux.
The sun does not stay still, nor does the moon
in orbit cease to wax or cease to wane,
nor do the other stars stay fixed in heaven;
they all move in a cycle, changing place.
At that time earth untrodden runs and the sun's orb,
as big as even now men have the power to vouch;
just so do all these things through one another race,
and, roaming, visit other places constantly;
we do not reach the middle place in union.
But whensoever Hatred to the vortex' utmost depths
arrives, and Love arises in the whirlwind's midst,
in Her then all these things unite to be just one.
Strive so my words reach not your ears alone,
and as you hear from me true facts take note:
I'll put before your eyes how She augments
with larger form life's union and increase,
and all that still remain of this creation,
first in the wild tribes of beasts that roam the hills,
then in the double race of man, then in the fruits
of rooted plants and grapes that mount the vine.
From these take in your mind proofs truthful of my words.

You'll see life's union and increase,
because of Hate, wreak deeds of separation.
This first the mass of mortal limbs makes clear;
at one time we unite in Love with all
the limbs that bodies have when life is blossoming;
another time again, split up by evil Strife,
they wander each apart amid life's breakers.
Just so it is for plants, and water-dwelling fish,
and beasts that sleep on hills, and fowls that mount on wings.

. . . . . that at life's end the limbs of all
fall from each other apart and meet their fate,
much though we will it not, from grim necessity,
as we decay. Though beauteous Love may hold us now,
the harpies with death's lots will come for us.
Woe that some ruthless day did not destroy me first,
before I used my claws to wreak dread deeds for food!
Now to no end my cheeks I wet with tears.
For we shall reach the unplumbed gyre, I fear,
and though men wish it not they'll have uncounted pains.
But we'll embark once more upon our tale.
When once the tireless flame did chance upon
all things, and caused their painful intermixture,
then creatures too progenitive were born
in every way, whose remnants still the dawn beholds.
When aither mixed did reach the utmost edge,
then birds flew up with shrieks and cries
tumultuous, but beasts, whose lot is lairs in earth
and grass, were born where earth was swathed around.
As when . . . .
(two damaged verses about a smith inlaying one metal on another)
First in the whelks with heavy backs that graze the sea
and in their stony mantles, then in oysters –
there you will see earth dwelling on the top of flesh.
Again, the armour-plate of strong-backed sea-crabs,
also the stone-skinned conches' and the tortoises' shells,
and spears of hornèd deer that roam the hills.
But listing all such creatures I'd not end.
As when . . . .,
from which . . . .,
and this . . . .,
. . . . .

(Here the papyrus breaks off)

Lucretius (94-?55 BC)

Titus Lucretius Carus was born in Samos, Greece and died in Athens. He was a philosopher, poet and author of an ethical philosophy of simple pleasure, friendship, and retirement, these being the hallmarks of epicureanism.

Lucretius

Lucretius founded schools of philosophy that survived directly until the 4th century AD. His six-book Latin hexameter poem ***De rerum natura*** variously translated *On the nature of things* and *On the nature of the universe*, survives virtually intact, although it is disputed whether he lived to put the finishing touches to it. As well as being a pioneering figure in the history of philosophical poetry, Lucretius has come to be our primary source of information on Epicurean physics, the official topic of his poem. Among numerous other Epicurean doctrines, the *atomic 'swerve'* is known to us mainly from Lucretius' account of it. He termed it *clinamen*. When atoms move straight down through the void by their own weight, they deflect a bit in space at a quite uncertain time and in uncertain places, just enough that you could say that their motion has changed. But if they were not in the habit of swerving, they would all fall straight down through the depths of the void, like drops of rain, and no collision would occur, nor would any blow be produced among the atoms. In that case, nature would never have produced anything.

This swerving, according to Lucretius, provided the *free will which living things throughout the world have.* Lucretius never gave the primary cause of the deflections. His defence of the Epicurean system was deftly and passionately argued, and was particularly admired for its eloquent critique of the fear of death.

De rerum natura is the best surviving explanation of the ideas of the Greek Epicurean philosophers. It describes the development of the cosmos, the Earth, living things, and human society through purely naturalistic mechanisms, without any reference to supernatural involvement. Among its aphorisms:

> What is food to one man is bitter poison to others.
> The fall of dropping water wears away the Stone.
> Life is one long struggle in the dark.

His six-book Latin hexameter *De rerum natura* included:

> O bountiful Venus, mother of the race of Aeneas, delight of gods and men, who, beneath the gliding constellations of heaven, fillest with life the ship-bearing sea and the fruit bearing earth; since by thy influence every kind of Comfort bring and mirth.
>
> To the ship-bearing seas, corn bearing earth.
>
> When the life of men lay foully grovelling before crushed beneath the weight of our eyes, a Religion, who displays her head from the region of the sky, lowering over mortals with terrible aspect, a man of Greece was the first that dared to raise mortal eyes against her, and first to make a stand against her. Him neither tales of gods or thunderbolts nor heaven itself with its threatening roar, repressed but roused the more the active energy of his soul, so that he should desire to be the first to break the close bars of nature's portals.

Accordingly the vivid force of his intellect prevailed, and proceeded far beyond the battlements of the world, and in mind and thought traversed the whole immensity of space; hence triumphant, he declared to us what can arise *into being* and what cannot; in fine in what ways the powers of all things are limited, and a deeply fixed boundary assigned to each. By which means religion, brought natural philosophy.

An early thinker in what grew to become the study of evolution, Lucretius believed nature experimented endlessly across the aeons, ***and the organisms that adapted best to their environment had the best chance of surviving.***

(This author's bold text, anticipating Darwin). Living organisms survived because of their strength, speed, or intellect. (However) In contrast to modern thought on the subject, (unlike Darwin) he did not believe that new species evolved from previously existing ones and denied that modern animals, which dwell on land, derived from marine ancestors. Lucretius challenged the assumption that humans are necessarily superior to animals, noting that mammalian mothers in the wild recognise and nurture their offspring as do human mothers.

Despite his advocacy of empiricism and his many correct conjectures about atomism (a natural philosophy proposing that the physical world is composed of fundamental indivisible components known as atoms), Lucretius concluded his first book stressing the absurdity of the (by then well-established) spherical Earth theory.

While Epicurus left open the possibility for free will by arguing for the uncertainty of the paths of atoms, Lucretius viewed the soul or mind as emerging from arrangements of distinct particles down under *our* feet.

Chapter 2

Buffon, Burnett, Diderot, Chambers, Matthew & Blyth

Georges-Louis Leclerc, Comte de Buffon (1707-1788)

Later in the 18th century, the French philosopher Buffon one of the leading naturalists of the time, suggested that what most people referred to as species were really just well-marked varieties, modified from an original form by environmental factors. For example, he believed that lions, tigers, leopards, and house cats might all have a common ancestor. He further speculated that the 200 or so species of mammals then known might have descended from as few as 38 original animal forms. Buffon's evolutionary ideas were limited; he believed each of the original forms had arisen **through spontaneous generation and that each was shaped by "internal moulds" that limited the amount of change**.

Comte de Buffon

Buffon's works, *Histoire naturelle* (1749-1789) and *Époques de la nature* (1778), containing well-developed theories about a completely materialistic origin for the Earth and his ideas questioning the fixity of species, were extremely influential.

James Burnett, Lord Monboddo (1714-1799)

James Burnett

Burnett was a Scottish judge, scholar of linguistic evolution, philosopher and deist. He analysed man's relation to other species and explored the origins of language and society and anticipated principles of Darwinian evolution.

He extended Rousseau's doctrine of the identity of species of Man and the chimp into the hypothesis of common descent of all the anthropoids, and suggested by implication a general law of evolution.

Charles Neaves, one of Monboddo's successors on the high court of Scotland, believed that proper credit was not given to Monboddo in evolutionary theory development. Neaves wrote in verse:

Though Darwin now proclaims the law
And spreads it far abroad, O!
The man that first the secret saw
Was honest old Monboddo.
The architect precedence takes
Of him that bears the hod, O!
So up and at them, Land of Cakes,
We'll vindicate Monboddo.

Erasmus Darwin noted his work in his publications in 1803. **Later writers have considered Monboddo's analysis as precursive to the theory of evolution.** Whether Charles Darwin read Monboddo is not certain. The latter debated with Buffon regarding Man's relationship to other primates. Charles Darwin did not mention Monboddo, but commented on Buffon:

> the first author who in modern times has treated [evolution] in a scientific spirit was Buffon

Buffon thought that Man was a species unrelated to lower primates, but Monboddo rejected Buffon's analysis and argued that the anthropoidal ape must be related to the species of Man: he sometimes referred to the anthropoidal ape as the "brother of Man". He suffered a setback, in his standing on evolutionary thought, because he stated at one time that men had caudal appendages (tails); some historians failed to take him very seriously after that remark, even though Monboddo was known to bait his critics with preposterous sayings.

In an updated genealogy of Burnett of the Leys family, Bailey stated that:

> Charles Darwin was to some degree influenced by the theories of Monboddo, who deserves the title of Evolutionist more than that of Eccentric.

He was a minor celebrity in Edinburgh because he was considered to be very eccentric. **But he actually came up with the idea that men may have evolved instead of being created by God. His views were dismissed because people thought he was mad, and in those days it was a very controversial view to hold. But he felt it was a logical possibility and it caused him a great deal of consternation. He actually did not want to believe the theory because he was a very religious person.**

He also elaborated on the advantages created by the adaptive change of primates to their environment and even to the evolving complexity of primate social structures. In 1772 in a letter to James Harris, Monboddo articulated that his **theory of language evolution** was simply a part of the manner that Man had advanced from the lower animals, a clear precedent of evolutionary thought. Furthermore, he established a detailed theory of how Man adaptively acquired language to cope better with his environment and social needs. He argued that the development of language was linked to a procession of events: first developing the use of tools, then social structures and finally language. This concept was quite striking for his era, because it departed from the classical religious thinking that Man was created instantaneously and language revealed by God. In fact, Monboddo was deeply religious and pointed out that the creation events were probably simply allegories and did not dispute that the universe was created by God. He was a vigorous opponent of other scientific thinking that philosophically questioned the role of God and was possibly the first person to associate language skills evolving from primates and continuing to evolve in early humans. He wrote about how the language capability had altered over time in the form not only of skills but physical form of the sound producing organs (mouth, vocal cords, tongue, throat), suggesting he had formed the **concept of evolutionary adaptive change**.

As an agriculturist and horse-breeder, he was quite aware of the significance of selective breeding and even transferred this breeding theory to communications he had with James Boswell in regard to selection of a mate. In his own works he stated that degenerative qualities could be inherited by successive generations and that by selective choice of mates, creatures could improve the next generation in a biological sense. This suggests that Monboddo understood the role of natural processes in evolution; artificial selection was the starting-point for many of the proto-evolutionary thinkers, and for Darwin himself.

He struggled with how to "get Man from an animal" (i.e. transmutation of species) without divine intervention. This is typical of the kind of thinking which is called *deist*. He developed an entire theory of language evolution around the Egyptian civilisation to assist in his understanding of how Man descended from animals, since he explained the flowering of language upon the spinoff of the Egyptians imparting language skills to other cultures. Monboddo cast early humans as wild, solitary, herbivorous quadrupeds. He believed that contemporary people suffered many diseases because they were removed from the environmental state of being unclothed and exposed to extreme swings in climate. He wrote of numerous cultures (mostly based upon accounts of explorers); for example, he described "insensibles" and "wood eaters" in *Of the Origin and Progress of Language*. He was fascinated by the nature of these peoples' language development and also how they fitted into the evolutionary scheme.

Against all this, Monboddo's contribution to evolution is today regarded by historians of evolution as being notable.

Charles Dickens knew of Monboddo and wrote in his novel, *Life and Adventures of Martin Chuzzlewit* about *"(...) the Monboddo doctrine touching the probability of the human race having once been monkeys"*. This is significant because *Martin Chuzzlewit* was published decades before Darwin wrote his theory of evolution.

Denis Diderot (1713-1784)

Denis Diderot was a French philosopher, art critic, and writer, best known for serving as co-founder, chief editor, and contributor to the Encyclopédie along with Jean le Rond d'Alembert. He was a prominent figure during the Age of Enlightenment.

Denis-Diderot

The ultimate freethinker, Diderot alternately flirted with deism, pantheism and atheism. Diderot's philosophy featured a mix of insights well ahead of their time – and nonsense. He was an early proponent of evolution, but unlike Darwin, he based his convictions more on thought experiments and readings than rigorous observation. Still, articles on fossils or "figured stones" almost certainly played a part in this thinking. Encyclopédie included articles on a fossil-rich German cavern, fossil shells, fossil bones, fossil mammoths, crystals and petrifactions. Influenced by the Roman poet Lucretius, Diderot concluded that organisms with organs well-suited to survival survived, and those with ill-suited organs simply did not. He disdained preformationism, realizing that reproduction was not a perfect process, that "mistakes" could occur in the process of reproduction and introduce new characteristics. He argued that environment could influence the traits of living things. Recognizing, along with Buffon, the overlapping characteristics between different species, ***Diderot believed that all species***

could be traced to a common ancestor. After initially rejecting spontaneous generation, he ultimately embraced the notion, again influenced by Lucretius. He believed that all matter was imbued with consciousness.

Diderot's quotes:

> Man will never be free until the last king is strangled with the entrails of the last priest.
>
> Only passions, great passions can elevate the soul to great things.
>
> The most dangerous madmen are those created by religion, and ... people whose aim is to disrupt society always know how to make good use of them on occasion.

After initially rejecting spontaneous generation, he ultimately embraced the notion, again influenced by Lucretius. He believed that all matter was imbued with consciousness.

Robert Chambers (1802-1871)

Chambers was a Scottish journalist. He was the author of *Vestiges of the Natural History of Creation*, an 1844 work of speculative natural history and philosophy. Published anonymously in England, it brought together various ideas of stellar evolution with the **progressive transmutation of species** in an accessible narrative which tied together numerous scientific theories of the age.

Robert Chambers (Courtesy of Peter Everett printsandephemera.com)

The work was initially well received by polite Victorian society and became an international bestseller, but its unorthodox themes contradicted the natural theology fashionable at the time and was reviled by clergymen – and subsequently by scientists who readily found fault with its amateurish deficiencies. The ideas in the book were favoured by Radicals, and its presentation remained popular with a much wider public. Prince Albert read it aloud to Queen Victoria in 1845. *Vestiges* caused a shift in popular opinion which – Charles Darwin believed – prepared the public mind for the scientific theories of evolution by natural selection which followed from the publication of *On the Origin of Species* in 1859.

For decades there was speculation about its authorship. The 12th edition, published in 1884, revealed officially that the author was Robert Chambers, a Scottish journalist, who had written the book in St Andrews between 1841 and 1844 while recovering from a psychiatric disturbance. Initially, Chambers had proposed the title *The Natural History of Creation*, but he was persuaded to revise the title in deference to the Scottish geologist James Hutton, who had remarked of the timeless aspect of geology: *no vestige of a beginning, no prospect of an end.* Some of the inspiration for the work derived from the Edinburgh Phrenological Society whose materialist influence reached a climax between 1825 and 1840. Phrenology (from Ancient Greek, meaning 'mind', and 'knowledge' is a pseudoscience which involves the measurement of bumps on the skull to predict mental traits). Chambers died in 1871 and was buried in the grounds of St Andrew's Cathedral, within the ancient chapel of St Regulus.

Among the early readers of *Vestiges*, Charles Darwin had conceived his own theory of natural selection to explain evolution six years earlier, and in July 1844 had written down his ideas in an 'Essay'. For a year he had been tentatively discussing his evolutionary ideas in correspondence with Joseph Dalton Hooker, who wrote to Darwin on 30th December 1844 that he had "been delighted with Vestiges, from the multiplicity of facts he brings together, though I do [not] agree with his conclusions at all, he must be a funny fellow: somehow the book looks more like a nine-days wonder than a lasting work: it certainly is 'filling at the price'. – I mean the price its reading costs, for it is dear enough otherwise; he has lots of errors." Darwin had read the book in November, finding that it drew on some of the lines of evidence he had been putting together, and introduced questions that had to be dealt with. He responded that he had been "somewhat less amused at it ... the writing & arrangement are certainly admirable, but his geology strikes me as bad, & his zoology far worse." Darwin had learnt geology from Adam Sedgwick, and was particularly interested in what his former mentor had to say about evolution. In October 1845 he wrote to his friend Charles Lyell that Sedgwick's review was a *grand piece of argument* ***against*** *the mutability of species* which he had read with "fear & trembling," but had been "well pleased to find" that he had anticipated Sedgwick's objections and "had not overlooked any of the arguments."

He read Explanations early in 1846 and thought "the spirit of [it], though not the facts, ought to shame Sedgwick", while noting speculation and evidence suggesting that Chambers had written the books. In April 1847, after meeting Chambers then subsequently receiving a presentation of *Vestiges*, Darwin became convinced that Chambers must have been the author.

In his introduction to *On the Origin of Species*, published in 1859, Darwin assumed that his readers were aware of *Vestiges*, and wrote identifying what he felt was one of its gravest deficiencies with regards to its theory of biological evolution:

> The author of the 'Vestiges of Creation' would, I presume, say that, after a certain unknown number of generations, some bird had given birth to a woodpecker, and some plant to the mistletoe, and that these had been produced perfect as we now see them; but this assumption seems to me to be no explanation, for it leaves the case of the co-adaptations of organic beings to each other and to their physical conditions of life, untouched and unexplained.

Chambers took the publication of the *Origin* as an opportunity to release a new edition of *Vestiges* and respond to Darwin's comments, lamenting that Darwin had misunderstood the *Vestiges*. "It seems to the author," Chambers wrote, "that Mr. Darwin has only been enabled by his infinitely superior knowledge to point out a principle in what may be called practical animal life, which appears capable of bringing about the modifications theoretically assumed in the earlier work. His book, in no essential respect, contradicts the present: on the contrary … it expresses substantially the same general ideas." In perhaps a gross simplification, Chambers concludes that "The difference seems to be in words, not in facts or effects." At the very least, Chambers saw in Darwin a much needed ally – one that he simply could not afford to have against him.

It is probable that Darwin read Chambers' comments, because he removed the offending passage from the 3rd edition of *Origin* (1861) and all subsequent editions. In a historical sketch, newly added to the 3rd edition, Darwin softened his language a bit:

> The author *(Chambers)* apparently believes that organisation progresses by sudden leaps, but that the effects produced by the conditions of life are gradual. He argues with much force on general grounds that species are not immutable productions. But I cannot see how the two supposed "impulses" account in a scientific sense for the numerous and beautiful co-adaptations which we see throughout nature; I cannot see that we thus gain any insight how, for instance, a woodpecker has become adapted to its peculiar habits of life. The work, from its powerful and brilliant style, though displaying in the earlier editions little accurate knowledge and a great want of scientific caution, immediately had a very wide circulation.

(I suppose this was because Chambers, being a journalist, knew how to write in a way that had appeal for the general public).

Darwin even suggested that Chambers' book helped pave the way for the publication of his theory of evolution by natural selection.

> In my opinion it has done excellent service in this country in calling attention to the subject, in removing prejudice, and in thus preparing the ground for the reception of analogous views.

The harsh reception that *Vestiges* received, and the mockery which was made of its evolutionary ideas, has been cited by some historians as a factor leading to Darwin's caution in publishing his own theory of evolution. In a letter to Thomas Henry Huxley in 1854 (five years before his own book on evolution was published but 12 years after its ideas had first been sketched out in an unpublished essay), Darwin expressed sympathy for the (still anonymous) author of *Vestiges* in the face of a savage review by Huxley:

> I must think that such a book, if it does no other good, spreads the taste for Natural Science. But I am perhaps no fair judge, for I am almost as unorthodox about species as the Vestiges itself, though I hope not quite so unphilosophical.

However, later the same year, in a letter to Hooker, Darwin mentioned *Vestiges* in a more sober tone:

> I should have less scruple in troubling you if I had any confidence what my work would turn out. Sometimes I think it will be good, at other times I really feel as much ashamed of myself as the author of the Vestiges ought to be of himself.

It was reading *Vestiges* in 1845 that first inclined Alfred Russel Wallace to believe that the transmutation of species occurred. It was this belief that would lead him to plan his early field work with the idea of collecting data on the geographic distribution of closely allied species in hopes of finding evidence to support the idea. Wallace made the following comments on the concept of transmutation of species as described in *Vestiges* in a letter to Henry Bates a few months after first reading it:

> I have a rather more favourable opinion of the 'Vestiges' than you appear to have. I do not consider it a hasty generalization, but rather as an ingenious hypothesis strongly supported by some striking facts and analogies, but which remains to be proved by more facts and the additional light which more research may throw upon the

problem. It furnishes a subject for every observer of nature to attend to; every fact he observes will make either for or against it, and it thus serves both as an incitement to the collection of facts, and an object to which they can be applied when collected.

Patrick Matthew (1790-1874)

Patrick Matthew

Patrick Matthew was a Scottish grain merchant, fruit farmer, forester, and landowner, who contributed to the understanding of horticulture, silviculture, and agriculture in general, with a focus on maintaining the British navy and feeding new colonies.

In 1831, writing *On Naval Timber and Arboriculture* he stated:

> Continual balancing of life to circumstance... (The) progeny of the same parents, under great differences of circumstance, might, in several generations, even become distinct (1790-1834) species, incapable of co-reproduction...

Darwin immediately wrote to his close friend, the geologist Charles Lyell, one of the scientists who had most influenced his idea of evolution:

> In last Saturday Gardeners' Chronicle, a Mr Patrick Matthews [sic] publishes a long extract from his work On Naval Timber & Arboriculture published in 1831, in which he **briefly but completely anticipates the theory of natural selection**. I have ordered the book, as some few passages are rather obscure but it, is certainly, I think, a complete but not developed anticipation! Erasmus always said that surely this would be shown to be the case someday. Anyhow one may be excused in not having discovered the fact in a work on naval timber.

And so he quickly responded in a letter to *The Gardener's Chronicle*:

> I freely acknowledge that Mr. Matthew has anticipated by many years the explanation which I have offered of the origin of species, under the name of natural selection. I think that no one will feel surprised that neither I, nor apparently any other naturalist, had heard of Mr. Matthew's views, considering how briefly they are given, and that they appeared in the appendix to a work on Naval Timber and Arboriculture. I can do no more than offer my apologies to Mr. Matthew for my entire ignorance of his publication. If another edition of my work is called for, I will insert a notice to the foregoing effect.

With this gentlemanly exchange of letters, perfectly documented, any possible controversy between the two was settled. And today Matthew, Darwin and Wallace are considered to be the only three people who discovered, each on their own, that natural selection is the mechanism of the evolution of species. It is also true that others had approached the idea (such as Thomas Malthus, who in his population studies of 1798 expressed it in a negative way, such as the struggle for existence or the competition for natural resources). James Hutton suggested in 1794 that natural selection led to improved varieties of species, and William Charles Wells proposed in 1813 that it could give rise to new varieties; in contrast, Edward Blyth concluded that this mechanism served just the opposite purpose – that of keeping species stable and unaltered.

Edward Blyth (1810-1873)

Edward Blyth was an English zoologist who worked for most of his life in India as a curator of zoology at the Museum of the Asiatic Society of India in Calcutta. He wrote three articles on variation, discussing the effects of artificial selection and describing the process in nature as restoring organisms in the wild to their archetype (rather than forming new species). However, he never actually used the term *natural selection*. These articles were published in *The Magazine of Natural History* between 1835 and 1837.

In February 1855, Charles Darwin seeking information on variations in domesticated animals of various countries, wrote to Blyth who was:

Edward Blyth

> much gratified to learn that a subject in which I have always felt the deepest interest has been undertaken by one so competent to treat of it in all its bearings

and they corresponded on the subject. Blyth was among the first to recognise the significance of Alfred Russel Wallace's paper *On the Law which has regulated the introduction of Species* and brought it to the notice of Darwin in a letter written in Calcutta on 8th December 1855:

> What think you of Wallace's paper? Good! Upon the whole! ... Wallace has, I think, put the matter well; and according to his theory, the various domestic races of animals have been fairly developed into species. ... A trump of a fact for friend Wallace to have hit upon!

There can be no doubt of Darwin's regard for Edward Blyth: in the first chapter of *On the Origin of Species* he wrote:

> *Mr. Blyth, whose opinion, from his large and varied stores of knowledge, I should value more than that of almost any one,* ...

And now to the Darwins.

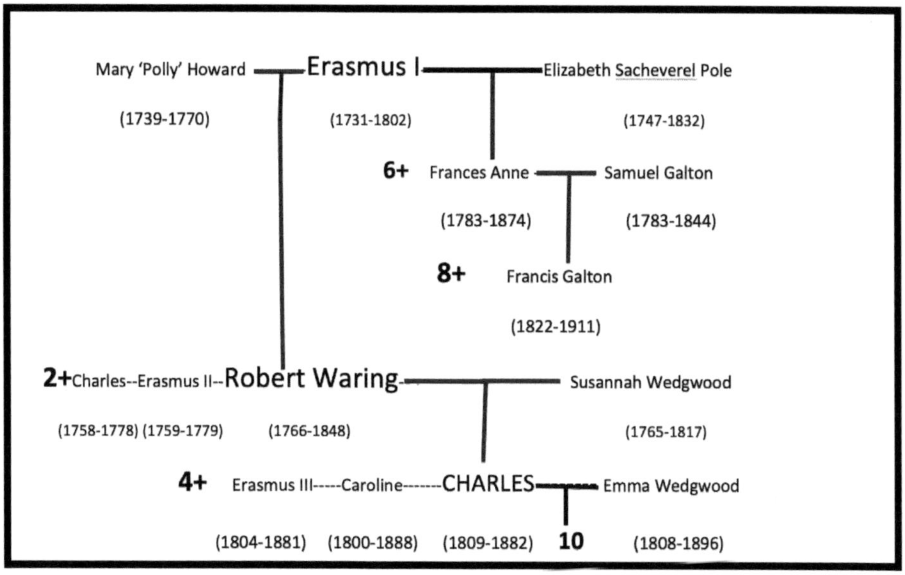

Abbreviated Darwin family tree

Bold numbers refer to total offspring numbers (Charles and Emma Wedgwood, 10), or additional siblings to those named, (Erasmus and Mary Howard, 2; Erasmus and Elizabeth Pole, 6; Frances and Samuel Galton, 8; Robert and Susannah Wedgwood, 4).

Chapter 3

Erasmus Darwin (1731-1802)

Erasmus Darwin was born on 12th December 1731 at Elston Hall in Nottinghamshire.

His father, Robert is credited with discovery of the first known fossilized skeleton of a Jurassic reptile in 1718. It was a part of a plesiosaur and formed part of the collection in the Natural History Museum in London. Erasmus had six siblings, three of each sex. From the age of 10, he was educated at Chesterfield School. His elder brother was keen on poetry and botany, an interest shared by Erasmus. After nine years at Chesterfield School, he matriculated, together with another brother, John, at St John's College, Cambridge in 1750. Robert had preceded them in 1743.

Elston Hall. From an engraving, pre-1754

His life at university was, unlike many of today's students, one of frugality and leaning towards temperance. He was especially interested in the classics, incorporating several themes in his poetry. Curiously, he decided it would be advantageous to be able to record lectures verbatim, and therefore he taught himself shorthand. 1752, the first year completed, lectures in medicine followed. These included those by **George Baker**, later to become a royal physician and discoverer that a puzzling illness, called *Devonshire colic* was due to lead poisoning.

For three months in 1753, he attended anatomy lectures by the famous **William Hunter**, in Covent Garden. Back in Cambridge he attended lectures by Noah Thomas, a royal physician, later knighted, and who in later times, proposed Erasmus Darwin to become a Fellow of the Royal Society, which was a great honour, and still is to this day. Also, he attended lectures by **William Heberden** (1710-1801), who became a famous and possibly immortal physician, whose description of osteoarthritis of the end finger joints is a teaching point to this day, as it typifies osteoarthritis, and not rheumatoid arthritis (although both diseases can co-exist in a single patient, along with many other co-morbidities). Heberden also described angina, the pain felt when the heart muscle is starved of blood. Additionally he wrote on the subject of the pulse, which provides many clues to heart disease.

In 1735, **Linnaeus** in his epic work, *Systema Naturae*, abandoning previous methods of classification of plants, focussed his attention on the flowers and, in a eureka moment of brilliance, realised that the wherewithal utilised by plants for reproduction involved two sets of structures which together, or by mingling with parts of other flowers of the same species, led to the production of seeds. These structures he bravely described as sexual, the female *pistil* and the male *stamens*. This presented a risk of causing offence in polite circles in England. **Thomas Martyn**, Professor of Botany at Cambridge, translated Rousseau's epistle into English as *Letters on the Elements of Botany Addressed to a Lady in 1785*. Martyn's target consisted of a particular class of enlightened British ladies interested in botany as an elegant and suitable pursuit i.e. lacking any hint of crudity. The letters elucidate the structure of plants and give their order in the Linnaean system. Rousseau avoided using Latin names so as to make the scientific content in his letters more accessible. The girls' tutor, Pierre Prévost, was highly appreciative of the letters, and commented: "Never has a botanist carried so far the delicacy and correctness with which he arranged the plants on paper … His book of mosses, in duodecimo format, was a little masterpiece of elegance." The letters remained unpublished at Rousseau's death; when they were finally published they received widespread acclaim. "It's a true pedagogical model, and it complements Emile," commented **Goethe**. According to Martyn, the letters were not meant for reading while sitting in an easy chair; rather they were directed at readers with a plant in their hand. **Dr Samantha George** a Senior Lecturer in Literature in the Social Sciences, Arts & Humanities Research Institute at the University of Hertfordshire is the author of a book on botany, sexuality and women's writing. She has stated that in the 18th century, many botanical texts were directed to women, and reproduction, sexuality, science, classification, solitude, and public debate were subjects confronting opinions on women's intellectual and moral faculties, and their social status. She wrote *Linnaeus in Letters and the Cultivation of the Female Mind: 'Botany in an English Dress.'*

Erasmus Darwin was also a talented poet in the making. In later years his *The Loves of the Plants* (1789) would form part of his epic, *The Botanic Garden*, published in 1791 which would earn him a place as a poet with serious talent. **James Lee**, in 1760, deduced from Linnaeus' Latin, that the stamens could be regarded as husbands, the pistils as wives, fertilisation 'sexual union' as marriage. Flowers lacking stamens were eunuchs! Or as Erasmus put it:

> Woo'd with long care, CURCUMA cold and shy
> Meets her fond husband with averted eye:
> Four beardless youths the obturate beauty move
> With soft attentions of Platonic love.

(Erasmus' explanation of Linnaeus:

> Curcuma. Turmeric. One male and one female inhabit this flower; but there besides four imperfect males, or filaments without anthers upon them, called by Linnaeus eunuchs).

With vain desires the pensive ALCEA burns,
And, like sad ELOISA, loves and mourns.
The freckled IRIS owns a fiercer flame,
And three unjealous husbands wed the dame.
CUPRESSUS dark disdains his husky bride,
One dome contains them, but two beds divide.
The proud OSYRIS flies his angry fair,
Two houses hold the fashionable pair.

William Withering, a Birmingham doctor, and member of the Lunar Society like Erasmus, tried to desexualise Linnaean terminology. He wrote:

> From an apprehension that 'Botany in an English dress' would become a favourite amusement with the ladies, many of whom are very considerable proficients in the study, in spite of difficulty; it was thought proper to drop the sexual distinctions in the titles to the Classes and Orders.

Consequently, he replaced the Linnaean *pistil* (female) with *pointal,* and *stamen* (male) with *chive.* Curiously the etymology of pointal is from old French *pointille ("a prick or prickle").* This is clearly at variance with the current vernacular! And chive is a vegetable renowned for possessing *long stamens! (One wonders if Withering got things somewhat mixed up!)*

Priscilla Wakefield, was an English Quaker philanthropist who wrote on feminist economics and scientific subjects, as well as producing children's non-fiction. In 1796, she published a *Series of Familiar Letters* between two imaginary sisters. She wrote:

> Till of late years, botany has been confined to the circle of the learned, which may be attributed to those books that treated of it, being principally written in Latin: a difficulty that deterred many, particularly the female sex, from attempting to obtain the knowledge of a science, thus defended, as it were, from their approach.

Richard Polwhele was a Cornish clergyman, poet who went to Christ Church, Oxford, continuing to write poetry, but left without taking a degree. In 1782 he was ordained a curate. He characterised botanic exploration as an uneasy blend of science and voyeurism. He couldn't understand how examination of a plant's organs of generation could equate to female modesty. He seems himself to have been quite taken by the whole business! He wrote, after observing children of both sexes quite innocently examining flowers:

> More eager for illicit knowledge pant,
> With lustful boys anatomise a plant;
> The virtues of its dust prolific speak,
> Or point its pistil with unblushing cheek.
> With bliss botanic as their bosoms heave,
> Still pluck forbidden fruit with mother Eve,
> For puberty in sighing florets pant,
> Dissect its organ of unhallow'd lust,
> And fondly gaze the titillating dust.

He actually stated:

> I have several times seen boys and girls botanising together.

Really!

However, Erasmus' medical studies were not yet completed, and in 1753, together with elder brother Robert, they set out by horseback to Edinburgh, then the best medical school. This was because most of the professors had spent time at the University of Leiden, where the most highly respected physician **Herman Boerhaave** taught. He was way ahead of his time, attempting to explain physical and chemical causes for all Man's maladies. Today we would invoke chemistry and haematology and rheology (or how fluids in motion behave). Erasmus stood out amongst his colleagues by revealing his knowledge of the classics, the result of his Cambridge experience. At Edinburgh he became friends with **James Keir** who later would play a prominent role with him in the formation of the **Lunar Society** in Birmingham. It is curious how medical thought evolves. The influence of Boerhaave was to the fore, but this did not invoke universal praise. Keir was one who felt that Boerhaave's ideas implied that Man was 'a hydraulic machine whose pipes were filled with fluids…'. Whilst there it appears Erasmus decided that sex and the consumption of alcohol were not essential or helpful in order to become a doctor, so he decided to cease drinking.

Curiously, when compared with the tight regulation of qualifications which form the part of any practising medical doctor today, there is no record of his obtaining a medical degree at Edinburgh, though he styled himself MD (according to **Desmond King-Hele**, in his book *Erasmus Darwin, a life of Unequalled Achievement*). **Anna Seward**, Erasmus' friend, states that he obtained the MB degree from St John's College in 1755 and that he defended his thesis 'that the movements of the heart and arteries are immediately produced by the stimulus of the blood.' This we now know is not the case, **William Harvey** in his work '*De motu cordis*' (which translated from Latin means 'On the movement of the heart'), in 1668 showed that the movement of the blood was due to the action of the heart and arteries, not the other way round. The heart is a muscle around a cavity, and when the muscle contracts, blood which has filled the cavity, is squeezed out forcefully, into the arteries, and around the body. It is curious that someone as mechanically minded as Erasmus did not realise that this was the case. Nothing deterred he started his medical practice in Nottingham in 1756. Whereas nowadays, any doctor employed in the National Health Service will have a surfeit of patients, in those days, although a good doctor would give of his time to heal the sick poor, that would not pay the bills. To succeed in attracting the more genteel classes, one needed letters of introduction, a fact that seems to have escaped him at the time. Nothing daunted, he turned his mind to improving the comfort of horse-drawn coaches, by inventing (at least on paper), sprung spokes. But then, out of the blue, a patient appeared. It was a man who, whilst drunk, was stabbed by another drunkard. Four days later, Darwin was called, and pulled out all the stops, (which included Peruvian bark, as used for malaria, and comprising quinine). Employing his training and medical education, he somehow conjectured that an abdominal stab wound might require treatment of a fever to save the patient's life, but to no avail, the patient died. Post-mortem ('dissection') revealed a pierced stomach, which probably had resulted in peritonitis, and death from sepsis. The peritonitis would undoubtedly have been the cause of the fever. He only stayed in Nottingham for two or three months. Undaunted, he decided to try his luck in Lichfield where he arrived in 1756 and where he was to remain for 14 years before moving to Derby for reasons which will become clear.

Home of Erasmus Darwin, Lichfield (Courtesy of Lichfield Civic Society)

He was better prepared for Lichfield than he had been in Nottingham. He bore letters of introduction to Lady Gresley (from his brother-in-law), and to the Reverend Thomas Seward, Canon at Lichfield Cathedral, who lived in the Cathedral Palace, an illustration of which is given opposite. Once in Lichfield, he would became acquainted with Seward's daughter Anna, aged 14, who later wrote a book, *On the Life of Erasmus Darwin*. She was 11 years his junior, and it might have been anticipated that as the years passed romance would flourish. If it did it was not apparent, though their correspondence may have been a sublimation of same. She wrote about his life mainly from the angle of his poetry. She was not medically trained, and possibly not medically interested being a literary critic. In this she faced an uphill struggle, for the recognition of women in this sphere was sadly lacking. They shared an interest in poetry though Darwin wasn't impressed with her romantic naivety. From Anna's romance aspect, it is possible that Darwin had been preceded by her harpsichord teacher, John Saville, whom she had met when aged 13. Their mutual interests included music, singing and dancing. However Saville married another and fathered two children.

Nothing daunted he remained a constant visitor to the Bishop's Palace incurring her father's order to cease the relationship in 1771. This was summarily ignored by the indomitable Anna Seward. And though clearly charmed by Erasmus, it is unlikely that his physical appearance made any positive contribution to their relationship. She wrote:

The Bishop's Palace, Lichfield (George Eksts, V & A images, with permission)

> He was somewhat above the middle size, his form athletic, and inclined to corpulence; his limbs too heavy for exact proportion. The traces of a severe smallpox, features, and countenance, which, when they were not animated by social pleasure, were rather saturnine than sprightly; a stoop in the shoulders, and the then professional appendage, a large full-bottomed wig, gave, at that early period of life, an appearance of nearly twice the years he bore. Florid health, and the earnest of good humour, a sunny smile, on entering a room, and on first accosting his friends, rendered, in his youth, that exterior agreeable, to which beauty and symmetry had not been propitious. Additionally he had a severe stammer.

In other words, ugly!

Lichfield Cathedral Church of the Blessed Virgin Mary and St Chad

Meanwhile, in the field of medicine, it must be asked, how could a medical doctor achieve fame and admiration when they knew very little if anything about the existence of germs, nor had ever seen bacteria through a microscope, nor knew anything about the immune system, nor had any antibiotics or understanding of the requirements of hygiene in order to avoid sepsis? The answer must be that most of their patients were probably not destined to die soon anyway, and that there was something about the manner of the doctor which evoked confidence and admiration. To quote Voltaire:

> The art of medicine consists of amusing the patient while nature cures the disease.

Although today's doctors have access to all the aforementioned, it must be remembered that medical practice relies both on the art, as well as the science. The science then was notable by its absence. Bleeding and purging was frequently the practice, but was based on fallacious theories, and was probably useless. The value of a knowledge of the blood pressure, though first measured in a tethered horse by the **Reverend Stephen Hales** in the 1730s, was not realised for over a century, as was the measurement of body temperature, or the use of the stethoscope.

However, Fortune played its part in apparently revealing Darwin's medical genius. A highly regarded local (Dr Wilkes) had attempted, but failed to cure a man suffering from a 'dangerous fever'. He had pronounced him as dying and beyond hope. (This actually was a 'get out' device adopted by many an anxious doctor whose reputation relied not only on curing, but also pronouncing when all hope was gone, and subsequently being proved right!). Accordingly, death was foretold by Dr Wilkes. Darwin's opinion was sought, and advice given. Somehow the patient recovered and Darwin's reputation was launched. Quite what Darwin did isn't known.

Darwin's enquiring mind revealed an interest in science and he further expanded his reputation by carrying out an experiment on the formation of clouds, dispelling the theory of one Henry Eeles of Dublin, who believed that electricity was involved. By a complex experiment, Darwin sought to prove him wrong, and sent a paper that was published in the prestigious *Philosophical Transactions of the Royal Society* in 1757. This provoked an angry response by Eeles who became increasingly frustrated when his arguments were not published, so that he was driven to collect all his protestations in a book called *Philosophical Essays*, published in 1771.

Darwin broadened his outlook in 1757 by falling in love with **Mary ('Polly') Howard** aged 17 who lived nearby. She was a friend of Anna Seward who thought her 'a blooming and lovely young lady'. Cupid's arrows as employed by Darwin consisted of verse. About a week before the wedding, whilst staying at Darlaston Manor with friends, he happened upon an old recipe book, several pages of which had been sampled by rats. Among the recipes was one named 'To make Love'. To his assumed disappointment, it was merely a long herbal. Another called 'To make a good Wife' resulted in his writing to Polly stating that he knew she would know better than anyone how to be a good wife, and anticipating her promise to this effect. His plan was to return to Lichfield from Darlaston Manor on a Wednesday in order to complete the paperwork and marry next morning at eight o'clock. His reason for this was to avoid being:

> more watch'd and teased by the Eye and Tongue of Impertinence.

As a respectable doctor, he was somewhat sensitive about being regarded as a cradle-snatcher. They married at St Mary's Church in Lichfield on 30th December 1757 and eight months later she presented him with a son, Charles. Anna Seward many years later made an observation on the marriage. She described Polly as, 'a blooming and lovely young lady of eighteen,' but felt that her health had suffered as a consequence of frequent childbirth. Polly died 13 years later, having given birth to five children two of whom also had died. Anna Seward went on:

> But alas! Upon her early youth, and a too delicate a constitution, the frequency of her maternal situation, during the first five years of her marriage, had probably a baneful effect.

We do not know what she died of. She mentions severe headaches and stomach pains, with complete periods of remission but only for a few days. Could it have been severe migraine, which has been associated with an increased risk of heart attack? Subarachnoid haemorrhage has rarely been reported with spinal, though not abdominal symptoms.

The West Midlands at this time was going through momentous changes that would usher in the Industrial Revolution. Erasmus was in contact with several intellectual and/or industrial giants, including **Joseph**

Soho House, Handsworth

Priestley, Josiah Wedgwood, Matthew Boulton, James Watt, Thomas Day, Richard Lovell Edgworth, Samuel Galton, James Keir, William Small, John Whitehurst, and William Withering. All were connected through the **Lunar Society**, whose members would meet when there was a full moon (hence the name), in order to facilitate their homeward journeys. They used to meet at Soho House, in Handsworth, Birmingham, the home of **Matthew Boulton**, as well as in Erasmus' house in Lichfield.

Shortly after their marriage, Erasmus and Polly moved to a large house in the cathedral close of Lichfield in 1758, and on 3rd September, she presented him with their first child. He was called Charles, and turned out to be an exceptional pupil and student of medicine. Tragically he was destined to die as a result of incurring an infection after dissecting a body in Edinburgh. Erasmus helped Anna Seward with her poetry, but made the mistake of teasingly mentioning to her father that her poetry was better than his (the father's). Seward Senior unfortunately took this the wrong way and was affronted. It led to her parents banning her from poetry, and instead encouraging more traditional feminine pursuits, namely needlework and music. Or did perhaps her parents feel that the brilliant already married doctor was getting uncomfortably close to their talented daughter?

The happy couple added a new front to their property in 1760. Erasmus was very busy, travelling up to 10,000 miles a year (King-Hele). This, considering there was no motor transport, is quite amazing, though admittedly there were no motorway delays, as we all experience nowadays. During these years, he established cordial relations with the Brummie cognoscenti, from the world of science and commerce. One such was **Matthew Boulton**, who was a successful manufacturer. He and Darwin together possessed the magnetic qualities of personality that were to the fore in the success of the Lunar Society, where all was not always sweetness and light. An example of this was **William Withering**, a Birmingham physician, who was to spectacularly fall out with Darwin's son Robert, a physician in Shrewsbury.

A teacher at Cambridge University and an earlier mentor of Darwin, was **John Michell** who would be regarded as a great seismologist from his observations on earthquakes (1760). But more amazingly, he foretold the existence of dark stars and black holes. He wondered what would happen if a star's gravity was so immense that the necessary velocity for light to escape would need to be more than the speed of light. Given that nothing can exceed the speed of light, this star would remain invisible since its light was trapped within it. He went on to become a rector at Thornhill, near Leeds. He introduced Darwin to **Benjamin Franklin** who carried out experiments on electricity and invented the lightning conductor. Franklin had founded the **American Philosophical Society**, which may have inspired Darwin to co-found its equivalent here, namely the Lunar Society. Other influential acquaintances included Samuel Garbett and John Roebuck, who constructed a sulphuric acid factory in Steelhouse Lane in Birmingham. Sulfuric acid is used in large quantities in iron and steel making principally as pickling acid used to remove oxidation, rust and scale.

The social life around the cathedral close centred on Darwin and the Sewards, comprising the Canon, Anna and her younger sister Sarah. Their conversation would tend to concentrate on art, poetry, music and literature, medical topics being a less likely contender. Imagine then Anna's feelings when Erasmus was offered the opportunity to dissect a human body! This would be available the day following the hanging of the miscreant. This was to take place on 25th October 1762, Darwin would dissect and lecture on its anatomy as long as the smell was tolerable.

In November 1763, Polly gave birth to a daughter, who unfortunately died four months later, of what we do not know. In 1764, Anna's sister Sarah became ill, and despite Dr Darwin's attempts, succumbed to what was presumed to be typhus, when her fever changed from 'inflammatory' to 'putrid'. (It seems unlikely that it was typhus, which is spread by body lice, and as such was common in jails, where closeness and poor hygiene were common. Another possible cause of a putrid fever was smallpox, which seems more likely).

Darwin's mind, ever creative, and not totally concerned with things medical, decided to improve transport. After all, we had Watt's steam engine, so why not apply the same principle, already applied to pumping water out from the coal mines, to the (optimistically) simple task of making the coach wheels go round? Thus we would, (if it worked), have the world's first steam car. He put the idea to Boulton, who unfortunately was not in a position to advance the necessary funds owing to his own debts, so it came to nothing. Another business interest was in creating canals, in particular the **Grand Trunk Canal**, eventually linking the Trent and the Mersey. This was obviously of interest to Wedgwood whose beautiful china was often sacrificed to the rutted roads of the time (what's changed with today's potholes?!) (These men had a vision, somewhat like today's of HS2 feeding the powerhouse of the North, or just as likely, and cynically, speeding the rush to the South!) On 21st April 1766, the Bill was passed in the House of Commons, Royal Assent was granted by George III in May, the first turf was cut by Wedgwood on 26th July. It was opened in 1771, and was 93 miles long.

Another revered and admired member was **William Small**, (1734-1775) born in Aberdeen. He attained a post as a physician in Birmingham, and somewhat displaced Darwin as Boulton's favourite physician. Darwin was not put out, obviously in thrall to the charm and acquired respect by all for Small. Small's partner in Birmingham was **John Ash**, (1723-1798) and together they collaborated in the coming about of the **Birmingham General Hospital**. His organisational skills were put to good use and the Lunar Society thrived as a consequence.

Erasmus Darwin was soon to meet the famous **Jean-Jacques Rousseau** (1712-1778), a renowned philosopher, composer and writer. He was born in Geneva, and his views on philosophy influenced the age of European Enlightenment, with ramifications in some aspects of the French Revolution, as well as modern political and educational thought. He and Darwin shared an interest in botany.

Another example of Darwin's genius was in the field of engineering. One problem with horse-drawn carriages at that time, was that they had a tendency to overturn when taking a sharp bend. The front wheels though mounted on a swivelling axle, would be obliged to maintain a parallel position to each other during the turn. However, as is apparent, the distance travelled by the inner wheel would be less than the outer, so resulting in a skidding motion. Erasmus Darwin's idea was to permit the axles to swivel independently so that neither wheel would skid, thus ensuring a lesser risk of overturning. The idea was subsequently patented in England by Rudolph Ackerman, having been re-invented by a Mr Lankensberger, of Munich, in 1818.

Ackerman Steering. In **Ackerman steering** each wheel is given its own pivot (which is typically close to the hub of the wheel). Tie rods create a trapezium shape with two additional pivots. As the shape is a trapezium, as the inside wheel turns, the outside wheel turns independently.

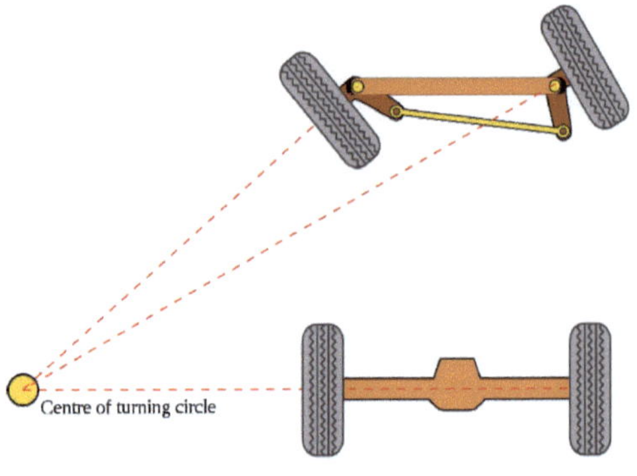

Ackerman steering (Wikipedia)

Presumably Darwin felt it would not help his reputation as a doctor to be tarnished by things so mechanical. But inventions no matter how good, frequently wither on the vine unless there is some financial backing. In Darwin's case it came in the person of **Richard Lovell Edgeworth**. Being a man of means, he was able to have constructed a Phaeton (a light four-wheeled open horse-drawn carriage) utilising Darwin's idea. This was communicated by Edgeworth to the Society for the Encouragement of Arts. Two more carriages were made and road tested. But Darwin's interest in road going vehicles was not limited to the horse-drawn variety. He felt that it might be possible to construct a steam driven carriage. The problem with steam engines at the time was that once the steam under pressure was released, and the piston driven, the water had to be heated again in order to achieve the necessary pressure. This was the weakness of the **Newcomen** engine, used to pump water out of the mines. The solution was the incorporation of a condenser to receive the steam, without having to start boiling the water again. And the inventor of this was **James Watt**. The meetings of these eminent men often took place in Lichfield at Darwin's house, and this led to the formation of the **Lunar Circle**, which evolved into the **Lunar Society**, to be based at Matthew Boulton's Soho House in Handsworth, Birmingham.

The coincidence in time and place of Darwin, Anna Seward, and the renowned **Samuel Johnson** (1709-1784) which potentially could have been an artistic and literary powerhouse, instead was an uneasy jealousy driven cauldron. Darwin was critical of Anna Seward's poetry, but at least it was better than her father's. Samuel Johnson was internationally famous, and a good orator. His opinions carried, at least in part on account of how they were presented, and to disagree was not invited. Darwin's views stood little chance in Johnson's company, partly because of his unfortunate stammer. Nevertheless seizing on Johnson's work on Shakespeare, in 1765, and Canon Seward's on the playwright Beaumont, Erasmus launched into verse thus:

From Lichfield famed two giant critics come,
Tremble, ye Poets! Hear them! 'Fe, Fo, Fum!'

By Seward's arm the mangled Beaumont bled,
And Johnson grinds poor Shakespeare's bones for bread.

Consequently it would be hard to imagine that Johnson would be expected to have any interest in cultivating any form of friendship with Darwin.

By 1766, Polly and Erasmus' first son Charles was seven, and Erasmus was six. Later that year their third son, **Robert Waring** was born. He was destined to be an exceptionally successful doctor in Shrewsbury, and the father of the most famous Darwin of all, Charles.

In 1768, Erasmus was involved in a road accident in which he was thrown from his carriage. He suffered a patellar (knee-cap) fracture which would be responsible for a permanent subsequent limp. Nowadays, such fractures, if displaced can easily be corrected by orthopaedic surgery to re-align the displaced fragments, but this was not available at that time. This seems a little odd, because the fracture is very superficial, and a displaced patella should be obvious to any doctor. If surgeons of the time could remove bladder stones, then surely an obviously out of line patella should be amenable to surgery, even if there was no anaesthetic. By a strange coincidence, Anna Seward suffered a similar injury the same year. It may be surmised that falling out of a carriage and suffering a fractured patella was common at the time, as perhaps nowadays, whiplash injuries are. But unfortunately, from 1776, Polly's health began to decline, which understandably had a profound effect on her loving husband.

However, his genius for invention was not daunted, and in this regard his horizontal windmill must be viewed. This, he claimed, generated one third more power than a traditional windmill of similar sail size. But Darwin's amazing ability, even genius for inventions, had to be kept in check, because his source of income was his medical practice. To be seen as an inventor of things associated with trades, like carriages, and windmills, could have a deleterious effect in a class-driven society. One example of his medical wisdom was when Wedgwood sought his advice in regard to spots he was seeing before his eyes. He understandably feared he might be going blind. Darwin from his experience knew that these symptoms were almost universal as people aged, and encouraged them not to look at them. We now know that the condition is what we nowadays term 'floaters'. These are small fragments of opacity within the eye's vitreous humour, and rarely cause any serious visual consequences.

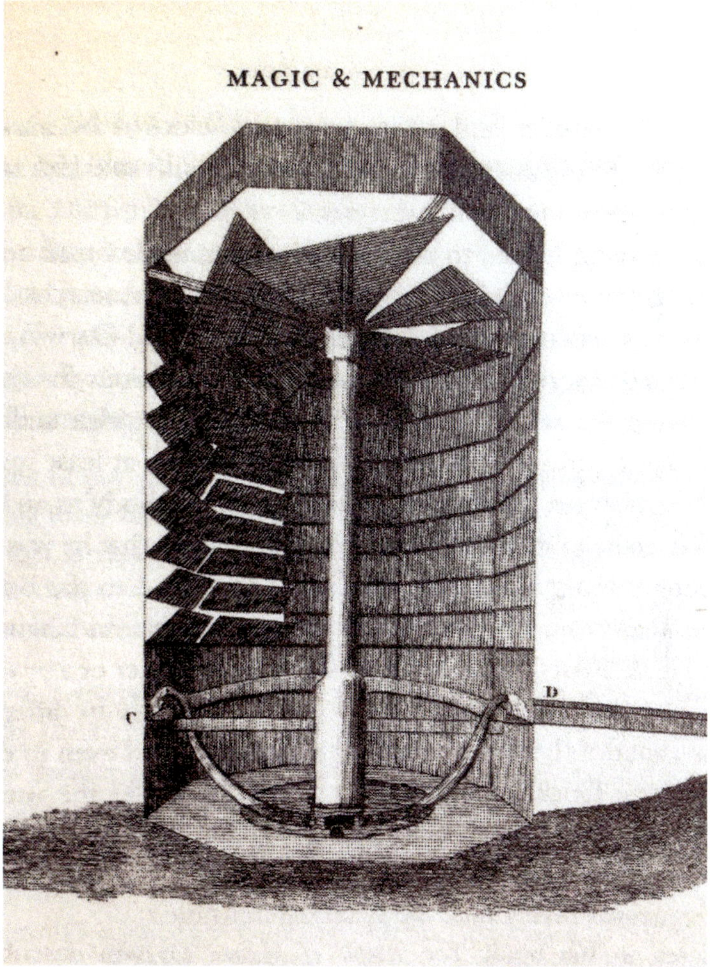

Erasmus Darwin's horizontal windmill (1768) (Twitter)

Coincidental to Polly's illness, whatever it was, Erasmus became less active socially, but his restless questing mind continued. It now began to focus on a more metaphysical plane, the interplay of mind and body prompted him into asking more fundamental questions about life, meaning, and the place, if any, of religion. Into this frame of mind, interposed itself the mysterious place of fossils. It would take a further 25 years before he put his thoughts into verse in ***Zoonomia***. It was one thing to surmise that fossil remains implied ancestors from which modern species had evolved, but another to conclude that there were common ancestors and that present day species had evolved from them. The fly in the ointment, was where God fitted in. And he knew he had to tread carefully. After all, most of his patients held conventional religious beliefs. He subscribed to the theory of common descent, namely that all life had evolved from a single microscopic ancestor. This idea led to the use of the term, **'theory of common descent'**. However at that time he still believed in the possibility that God had started it, and that from a 'living filament' all else evolved. He devised a way of expressing his excitement and pride at this eureka moment, by adding the Latin '*e conchis omnia*' to the family motto, and had it affixed to his carriage in 1770. It meant literally, 'everything from shells.' So here we had the ebullient Erasmus, escaping from the growing anxiety engendered by Polly's ill health, advertising his amazing thoughts emblazoned on his carriage based in the precinct of the cathedral, not a stone's throw from Canon Seward. As might be expected, the Canon was furious. The Greek philosopher Epicurus had made an impact on religious thinking by his philosophy of life, which was not overly, if at all, concerned with an afterlife. Seward compared Darwin in a satirical, one might say cutting verse:

> He too renounces his Creator,
> And forms all sense from senseless matter.
> Great wizard he! By magic spells
> Can all things raise from cockle shells…

Darwin sensed that he was heading for trouble with entrenched religious beliefs, and decided to remove the aforementioned appendage to the family motto. Anna Seward was by this time aware that the friendship of Erasmus and her father was now at an end.

In 1770 tragedy struck with the death of Polly, aged 30 after a year of severe suffering. She clearly suffered a lot of pain in her illness, requiring frequent doses of morphine, and became quite delirious towards the end. As a doctor, I find it intriguing that Dr Darwin conceded not one clue, as to the nature of his wife's illness, except that she suffered great pain. Her terminal delirium was most probably the consequence of her medication. On 4th July, Polly was buried in Lichfield Cathedral. She had brought up her offspring with a strong sense of morality, devoid of religious bias (or superstition as Erasmus would have it). This paved the way for acceptance of his evolutionary thinking which emerged years later (1794) in his work ***Zoonomia***. However 21 years later their son Robert, by then a physician based in Shrewsbury, being worried lest his mother's illness be hereditary probed his father into revealing more details. Though there has been some doubt about which side of her abdomen the pain was felt, Erasmus said:

> She was seized with pain on the right side of the lower edge of the liver, followed by violent convulsions...

This suggested gall bladder disease, probably with a stone impacted in the common bile duct. To quote from Bailey's *Text Book of Surgery*, 1962:

> Suddenly the patient experiences excruciating pain in the epigastrium and right hypochondrium – in severe cases it 'doubles her up' and she rolls in agony on the floor...

This book explains that morphine should never be given in these cases, because it increases muscle spasm, and pethidine is preferred, but this wasn't available until 1939. It was perhaps unusual that she never showed any signs of jaundice.

Aged 38 years, Erasmus was left to look after Charles (11yrs), Erasmus (10yrs) and Robert (4yrs). The doctor's sister, Susannah aged 40yrs stepped into the breach. It seems, though it is a little vague, that Charles and Erasmus Junior were at boarding school in London, but returned in time to see their mother before she died. Erasmus needed someone in addition to his sister to look after Robert. Consequently a young girl called Mary Parker aged 17yrs came from Elston to look after Robert. With his children thus looked after, and now a widower, Erasmus began writing his tome dedicated to medicine, and animals, namely the aforementioned *Zoonomia*.

In 1770 the artist **Joseph Wright** painted Erasmus' portrait, as well as that of the newly arrived **Thomas Day**.

Erasmus Darwin, (1731-1802) portrait by Joseph Wright (Wikipedia commons)

He was one member of the Rousseau, Edgeworth and Day triumvirate. Anna Seward and her circle were about to witness the attempt by Day to marry. But not to anyone. He had decided that he would first find, then mould a young girl to his design. By this he meant, a girl devoid of airs and graces, and with Rousseau's ideas on education in his mind, he set about the process. This would doubtless have aroused violent protest in society today, with the unfortunate events that have plagued several orphanages. He visited two of such, armed with testimonials of moral probity from a lawyer called John Bicknell. Thus prepared, he procured two 12yr old girls, of considerable beauty. He took them to France, in order to perfect the process. But his plan went awry, as they led him a merry dance, and he quickly disposed of one, to the employ of a milliner's. The other, whom he had named Sabrina, he brought back to Lichfield where she became quite the focus of attention before being packed off to boarding school. Later, duly educated she returned on occasion to Lichfield where she was regarded as a celebrity. The aforementioned lawyer hadn't forgotten his part in the process, and eventually married her himself, leaving Thomas Day to his bachelorhood.

Darwin's questing mind meanwhile was not idle. He became imbued with the idea of canals, and in particular with a narrow gauge canal for

Lichfield. Standard larger canals were far too expensive. He attempted to inspire his friends Boulton and Wedgwood, but to no avail. By contrast, the opposite would prevail on the railways where Isambard Kingdom Brunel's vision for the Great Western wide gauge railway would be passed over in preference to the standard (narrower) gauge now still adopted in most of Europe and Britain. Next he turned his genius to the production of a **speaking machine**.

> I contrived a wooden mouth with lips of soft leather, and with a valve at the back part of it for nostrils, both which could be quickly opened or closed by the pressure of the fingers, the vocality was given by a silk ribbon about an inch long and a quarter of an inch wide stretched between two bits of smooth wood a little hollowed; so that when a gentle current of air from bellows was blown on the edge of the ribbon, it gave an agreeable tone, as it vibrated between the wooden sides, much like a human voice. This head pronounced the p, b, m, and the vowel a, with so great nicety as to deceive all who heard it unseen, when it pronounced the words mama, papa, map, pam; and had a most plaintive tone, when the lips were gradually closed. My other occupations prevented me from proceeding in the further construction of this machine; which might have required but 13 movements, as shown in the above analysis, unless some variety of musical note was to be added to the vocality produced in the larynx; all of which movements might communicate with the keys of a harpsichord or forte piano, and perform the song as well as the accompaniment; or which if built in a gigantic form, might speak so loud as to command an army or instruct a crowd.

Erasmus Darwin's speaking machine (Wikimedia Commons)

Quite of what practical use this could ever be was addressed by Darwin. He thought it might be of use to address a crowd. This seems reasonable much like a modern day loudspeaker.

But his questing was not entirely occupied by inventing, and healing the sick. He was sharing his household with his sister Susannah, and Mary Parker, now 18yrs. He was a sad widower, whose body may well have been as fit as his mind. So it appears, he took a shine to Mary, who eventually bore him two daughters, to be known as the 'Miss Parkers', who were brought up in his home for 20 years. A rather odd arrangement one would have thought, the man a widower, his elder sister helping, and now a young pretty thing whom he never married and who bore him two daughters. And what of Anna Seward? It would appear that though Anna possessed charm and intelligence in spades, Erasmus wanted a young woman to satisfy his needs, and Anna was not such, despite her cultured appearance and manner. Instead, she had a warm but celibate relationship with a married man, John Saville, a fine choir singer.

Erasmus meanwhile was preoccupied with tending to the sick Mrs Wedgwood. During 1772, he had to make several 60 mile round trips to attend her. The nature of her illness is uncertain, but it included episodic dizziness and vomiting. It was not serious because she survived more than 40 years. It may have been a chronic vestibular neuronitis, a condition involving the balance mechanism of the ear, and characterised by the aforementioned symptoms. In order to make ends meet, Darwin had to tend the wealthy sick, but he also gave of his time to the poor. Travel was naturally an exhausting business, even more than nowadays with our overburdened highways. But in addition to medicine, as has been described, he had spent time devising a speaking machine, a horizontal windmill, a novel design for steering carriages, and also a method of springing coach wheels by making the spokes sprung, plus a device for duplicating handwriting, and of course his poetry. To this he now added an interest in chemistry.

The leading name of the time was that of the Frenchman, **Antoine Lavoisier**, (1743-1794), who eventually would be executed by guillotine on trumped up charges of stealing from the treasury. Lavoisier made important experiments on combustion, naming oxygen and hydrogen, and refuting the 'phlogiston' theory. Another was **Joseph Priestley**, (1735-1804), who discovered oxygen. He lived in Leeds, but moved to the Fairhill area of Birmingham, (nowadays Sparkbrook). He was a member of the Lunar Society.

There, a year after the French Revolution, his house was razed by a mob so he left Britain for the United States. After the French had had their revolution, it was the turn of the Americans, who with some justification objected to the attitude of the British politicians, who remained hell bent on continuing their exploitation of the colonies. Taxation without representation was the Americans' objection. The Boston Tea Party took place in December 1773.

Erasmus' three sons provide an interesting contrast in personalities and abilities. It was clear from the start that Charles was brilliant, and studious. He held his father's admiration. Next was Erasmus, who though undoubtedly intelligent, was somewhat diffident, and more inclined to having more literary interests. In this he was accompanied by Anna Seward. The third son, Robert Waring receives little mention in the book written by Desmond King-Hele on Erasmus Darwin's life. Into this literary circle entered Brooke Boothby, who had befriended Rousseau, and was a staunch disciple of his. Another visitor in 1774 was **Samuel Johnson**, (1735-1752), undoubtedly the more famous of the two, and one gathers there was a rather cool respect between the two.

In 1777, **James Watt**, whose major contribution was the steam engine, settled in Birmingham, and worked with Matthew Boulton who had acquired the rights to the engine as a consequence of Roebuck's financial difficulties in Scotland. Boulton employed nearly a thousand men in his 'manufactury' works at Handsworth in Birmingham. The axis of excellence in the field of engineering was now firmly entrenched in Birmingham, Lichfield was more for literary and fanciful occupations. Erasmus for whatever reason had lost interest in Mary Parker, and had once more become 'available' to the fair sex. He was on the prowl!

Meanwhile, with the gradual migration of intellectual excellence from Lichfield to the nascent Lunar Circle in Birmingham, a major blow was the death of perhaps its most deeply loved member, the kind and highly regarded **Dr William Small**.

Dr William Small (1734-1775), portrait by Tilly Kettle (Wikipedia)

Following his death, it was decided by the members to formalise the meetings, the first of the newly named Lunar Society being held on 31st December, 1775. The intention was to hold meetings at the full moon to expedite night time travel at the end of the meeting. Darwin wished to find a suitable successor to William Small. He had previously met a young doctor, who was born in Wellington, Shropshire, an Edinburgh graduate, whose first job was at the Stafford Infirmary. However, that job was not particularly lucrative, and therefore Darwin's offer to recommend him for the now vacant post in Birmingham was an attractive one, earning about £600 a year.

Dr William Withering, (1741-1799), portrait by Carl von Breda (art.uk)

William Withering was duly appointed to work with the senior physician **John Ash** at the General Hospital. But trouble was brewing, because Darwin, when he learned that Withering was writing an English version of the major botanical bible, employing the Linnaean classification, assumed Withering would be happy to co-author with him. Nothing could be further from the truth! So the seed was sown for their animosity. However, it would be several years later that a major row between them,

Carl Linnaeus, (1707-1778). Portrait by Alexander Roslin (Wikipedia)

and Erasmus' son, Dr Robert Darwin, by that time a doctor in Shrewsbury would break out. However matters would not come to a head for a while, because Erasmus was living and practising in Lichfield, Withering in Birmingham and later, Robert in Shrewsbury.

Linnaeus (1707-1778) classified plants in his most famous work *Genera Plantarum.*

He recognised in the plant world, that sex played the same role as in the animal world. Pistils were the female organ, and stamens the male, see below. Pollination was their sexual contact, leading to fertilisation, and seed formation. His work was in Latin, Withering translated it into English, but felt that the word 'pistil' was too suggestive for English ladies, and instead coined the word 'pointal'. Likewise, stamen became 'chive'. Prudery was commonplace, but that was on the surface. Beneath, sex and illegitimacy were alive and thriving. Especially amongst society. However, ladies in society were able to sublimate these thoughts by cultivating their interest in botany.

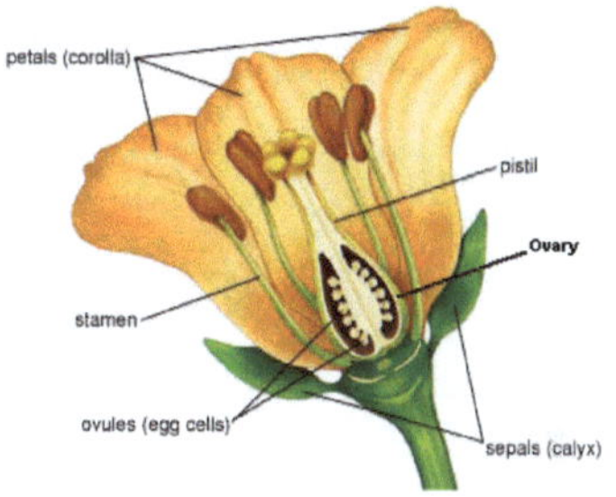

(Courtesy of Amrita Learning, Kochi, India)

A significant event took place in 1778. **Elizabeth Pole**, the wife of Colonel Pole (1717-1780) from Radburn in Derbyshire, had children suffering from whooping cough. She sought Erasmus' advice. After meeting her, the good doctor couldn't help being entranced by this young beautiful woman, married to a considerably older man.

In 1775 Charles, Erasmus' most admired son, and whose father wished him to become a medical doctor, entered Christ Church in Oxford, but was soon disenchanted at the emphasis on the classics, preferring to pursue medicine at Edinburgh. At the same time, Erasmus, the older son of

Erasmus, was studying law in London. In 1776, Erasmus Senior wrote a manuscript on the art of shorthand. However, he was ever the doctor, and that year he began his *Commonplace Book* comprised of medical histories, as well as his ingenious inventions. This book is housed in the Darwin Museum at **Down House**, which was to be Charles, Erasmus Senior's grandson's, home. Perhaps it would be more appropriately accommodated in the Erasmus Darwin Museum in Lichfield? Its case notes cover a motley collection of medical conditions, but one in particular deserves attention. It concerns a woman with asthma, and 'the dropsy' (meaning accumulation of body fluid in the tissues, nowadays termed 'oedema'.) He prescribed four draughts of a decoction of foxglove, following which she suffered diarrhoea and vomiting, and sudden death. It will be of interest that the world's expert on the use of the foxglove leaf for the treatment of the dropsy, was none other than William Withering, who had published his findings, in 1785 *An account of the foxglove and some of its medical uses* emphasising how to gradually arrive at the correct dose by judicious trial and error. Contrast this with Darwin's reckless manner of its prescribing. No wonder these men were at daggers drawn, and soon would be even more so after the **Houlston** case, to be described later.

Interestingly, in 1776 Darwin manifested a subtle change in his priorities. Whereas up to now, the emphasis was on mechanical invention both by Darwin himself, and his Lunar friends, it may well be that now he turned towards botany and gardens, having been excited by the Linnaean sexualisation of botany and seeing it as a means of expressing his emotions, particularly with Elizabeth Pole in mind. Also, he felt that it provided an escape from the world of commerce, so much to the

The Anderton boat lift (with permission from DavidDixon@btopenworld.com)

fore in Birmingham at the time. This would lead to his most famous poem, *The Botanic Garden*, which influenced Wordsworth and Coleridge. In addition, and by no means a coincidence, Elizabeth Pole was an obsessive gardener. However, lest one think Darwin had by now turned his back on all things mechanical, nothing could be further from the truth. He invented a canal lift, whereby boats instead of as hitherto using a conventional lock, instead would be driven onto a larger vessel, which would then be raised or lowered by a system of pulleys. Examples of this invention are still to be seen e.g. the Anderton boat lift in Cheshire.

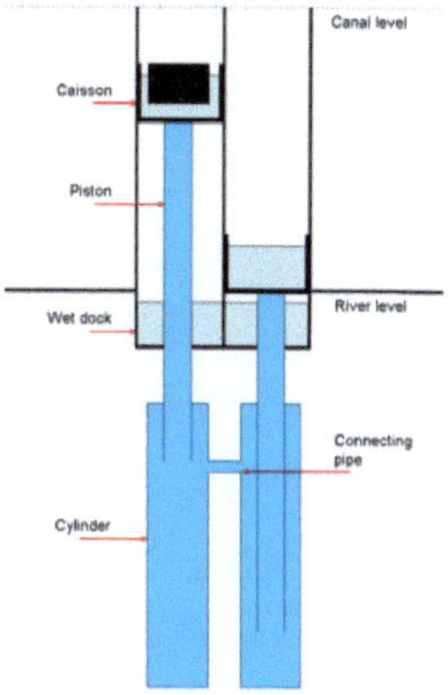

The Anderton boat lift

He also invented an artificial bird with flapping wings driven by a watch spring! To quote:

> Before flight begins, the wing lies folded back against the body. The cycle is then (1) out from the body, the leading edge moving radially forward about point x, like an opening fan; (2) down, while fully open; (3) back and in towards the body to fold the wing; (4) up, in closed position.

This has been described as original, no-one previously having described this sequence either for Nature's method, or of an artificial one. **Clive Hart** (1931-2016), author, concluded that Darwin's bird was '*the first, after two thousand years of speculation, to have been described with a complete, if summary, account of a power plant and the intended cycle for the wings. As such it deserves a special place in the history of aviation.*'

Darwin's pursuit of Elizabeth Pole was still annoyingly frustrated by her husband, but chance played into Darwin's hand when little Milly, aged 3yrs fell ill. She needed to be treated as an in-patient at Darwin's house in Lichfield, duly attending along with her mum, and a chaperone. The treatment (consisting of opium!), lasted a few weeks, and several of

Elizabeth Pole's children were affected. It may have been erysipelas, which can result in the striking red rash which was a feature, but against it was the apparent lack of fever.

The year 1778 would never be forgotten by the family. Tragedy was soon to occur, involving the brilliant medical student Charles, in Edinburgh. There, whilst dissecting a dead baby that had died of a brain infection, and not wearing surgical gloves, he cut his finger. Consequently he contracted a fatal infection, despite the best efforts of all the doctors, including his father who had hurried to Edinburgh.

Erasmus composed a poem based on the tragedy including:

> From birth I knew him; ev'n his infant plays
> Created love, awak'd the voice of praise;
> And as he grew, attracted by her charms,
> He woo'd fair Science, won her to his arms.

He also turned his hand to weather forecasting using a weather vane whose spindle reached down inside his house. He realised that certain changes in wind direction predicted change, and thus without having to venture forth out of doors, he had an idea of what the weather had in store for him before he set out to visit a patient many miles away. And one such in the autumn of 1778 was the aforementioned Elizabeth Pole. Eager Erasmus felt that he ought to stay the night in order to keep her under observation, but Colonel Pole thought otherwise. Nevertheless Erasmus spent the night under a tree beneath her window! She recovered, and probably would have anyway.

Darwin's next adventure was the resuscitation of his horizontal windmill in 1779. It was developed and installed at Etruria (today a suburb of Stoke on Trent) and used for grinding colours for use in pottery. Later that year, **Joseph Priestley**, the great experimental chemist arrived in Birmingham. He owned a large house in Fairhill (nowadays Sparkbrook), but which later that year was burned down by a rioting mob. Withering's house in Harborne (nowadays the mansion of the Edgbaston Golf Club) was also attacked, but Withering had cleverly arranged protection by means of plying the mob with alcohol, and having hidden his valuables in the nearby Edgbaston Old Church.

Radburn Hall (Wikipedia)

Dr Darwin's pathway to Elizabeth Pole was cleared when, on 27th November 1780, the good Colonel passed away, aged 63yrs. One assumes Darwin played no part in this, there is no mention of any intervention by him. Darwin had courted her through verse. Now, on the 28th he travelled to the Poles' house to pay his respects, knowing that it was now crunch time. How would Elizabeth, aged 33yrs regard him, an obese, pock-marked stammering man of 49yrs? What chance had he when pitted against the young bloods, engaged in fox-hunting, or military matters?

Quite against the run of the play, she chose Darwin, but on one condition. They must not live in Lichfield. Why? Perhaps she felt uneasy at the proximity of Anna Seward. They would have to live in her home, Radburn Hall, four miles west of Derby.

It may have been because Elizabeth disliked either the religious atmosphere of the cathedral close in Lichfield, or else, the proximity of Anna Seward's clique. In addition one wonders whether Mrs Pole played this card in order to test the seriousness of her suitor. But one needs to remember, the fires of passion burned with fierce intensity in Erasmus, so he agreed to move, and thus married Elizabeth Pole on 6th March 1781. Having achieved the desired prize, he swore in verse that he would forever remain faithful. In the event she was to bear him seven children, in addition to her previous three, (plus one who had died in infancy). For Darwin, life in Derby virtually ended his Lichfield and Birmingham based social life. Of course the Lunar Society felt a sense

Anna Seward (1742–1809), portrait by George Romney (With permission University of Vermont)

Elizabeth Sacheverell Pole and baby. Portrait by Joseph Wright, Derby (Courtesy of Lord and Lady Chichester)

of deprivation. He was always so full of new ideas, and arguments for or against those of others, a real live wire. One of these, ultimately to be proved true, concerned the composition of water. Darwin speculated that water was composed of a gas which could be displaced by the action of oil of vitriol (sulphuric acid), from 'its earth' (a metal). This gas is hydrogen. Subsequently Joseph Priestley showed that by passing an electric spark through a gas jar containing hydrogen and air, the glass surface became wetted. The wet substance was of course water. Previously it had been thought that water was an element that could not be broken down.

Erasmus Darwin set himself the gargantuan task of translating the Linnaean classification of plants into English. Its first instalment of 176 pages, appeared in 1782, titled, *A System of Vegetables, according to their classes, orders* – translated from the *Systema Vegetabilium* of Linnaeus by a Botanical Society at Lichfield. Darwin thanks 33 other botanists for their contribution, deliberately omitting Withering. He goes on:

> Dr Withering has given a Flora Anglica under the title of Botanical Arrangements ... but has entirely omitted the sexual distinctions

> which are essential to the philosophy of the system; and has ... rendered many parts of his work unintelligible to the Latin Botanist.

The omission of sexual distinctions was not the case. It was just that Withering was perhaps somewhat prudish, preferring to call the male components i.e. stamens, pointals, and the female pistil, chive.

In all Darwin's work amounted to 1,000 pages. A magnum opus, intending *en route*, to blow Withering's version out of the water. Withering's was first published in 1782, and went on for more than five editions, the fifth comprising approximately 2,000 pages, and edited by William Withering F.L.S Junior. No wonder the animosity between the two was destined soon to reach fever pitch.

The marriage with Elizabeth was a happy one, and was complemented by the birth of Edward in January 1782; Violetta, April 1783; Emma, August 1784; Francis (son), June 1786; John, September 1787; Henry, April 1789, died 1790; daughter Harriet, in July 1790. Also, a house was purchased in Derby as it would be more convenient in the winter. For this house, he designed a water closet! The move to Derby from Lichfield had two effects, firstly, it put him at a greater distance from Birmingham, and the Lunar Society meetings at Soho House in Handsworth, and secondly, he cut himself off from Lichfield. The latter naturally hurt Anna. Professionally however, he was able to continue his medical practice with the locals.

His inventiveness refused to lie down. He devised a means of opening and closing his greenhouse window using a see-saw mechanism which could see, or saw depending on the pressure applied to a bar by an expanding hydrogen filled globe in warm weather, or cooling mercury when the opposite prevailed.

He also envisaged travel by balloon, the Frenchman **Montgolfier** and his brother having shown the way with a hot air balloon, and a hydrogen balloon in Paris by **J. Charles** in 1783. Whereas these intrepid chaps travelled precariously by balloon, Darwin did so in verse, visiting planets and stars *en route*:

Rise, great Montgolfier! Urge thy venturous flight
High o'er the Moon's pale ice-reflected light;
High o'er the pearly Star, whose beamy horn
Hangs in the east, gay harbinger of morn;
Leave the red eye of Mars on rapid wing,
Jove's silver guards, and Saturn's crystal ring;
Leave the fair beams, which, issuing from afar,
Play with new lustres round the Georgian star…

Darwin formed the **Philosophical Society of Derby**, c.1783, with meetings taking place every Saturday evening in each other's houses. However, they were not as well attended as those of the Lunar Society despite Erasmus' enthusiasm. In the same year, there was an outbreak of some undefined disease in local cattle. He employed the classical technique of contact tracing, and concluded by advising the slaughter of all infected cattle at the onset of symptoms of the illness, and their timely burial. This was still not quite as thorough as the British policy of slaughtering all cattle of affected farms, even before any signs or symptoms became apparent in the case of foot and mouth, or 'mad cow' disease.

Radburn Hall was the Darwin's formal home, but bearing in mind his medical practice, it was decided to purchase a city residence in Derby itself.

Therefore he purchased a three-storey terraced house in Full Street, see picture. It possessed a walled garden backing on to the River Derwent. Unfortunately the river water was undrinkable because of industrial effluent from the nearby silk mill. Not daunted he had constructed an artesian well, basing the idea on his geological knowledge. 'Many mountains bear incontestable marks of their having been forcibly

Full Street, Derby

raised by some power beneath them – .' The term 'artesian' was not coined until the 1830 after a well in the Artois region of France. Perhaps it would have been more appropriately named a 'Darwinian well'. Not satisfied with this applied technology, he turned his attention to aeronautics, and in particular, ballooning using hydrogen, produced by reacting iron filings with sulphuric acid diluted 1:6 with water. It flew (fortunately unmanned) for 30 miles!

In July 1784, he gave the Presidential address to the Derby Philosophical Society gathered at his town house. He spoke about the means whereby science evolved utilising the special senses with mathematical, geometric and motion constructs. Communication was revolutionised by use of the alphabet, (a forerunner of the information technology of today). In the same year, his philanthropic qualities were revealed, when he began a project consisting of the provision of a dispensary for poor sick people. Later, much later, an infirmary was built on the initiative of **William Strutt**, in 1810. The health of the public was foremost in his mind, and in particular, the threats to health by alcoholic excess. One of his patients, no less than the Duke of Devonshire, who lived at Chatsworth House, was the recipient of some, not very welcome advice:

> I express myself strongly on this head, as I know in this kind of disease Health is not to be bought at any other price.

He went on to claim that alcohol excess was much worse than gluttony and stated that:

> when this sugar is converted into spirit by fermentation … our food is converted into poison; and becomes the greatest curse of the Christian world, destroying more than the sword, and wisely interdicted by the disciples of Mahomet and Confucius.

A totally different side of his nature, was his love of poetry. He was destined to be one of our leading poets, highly respected, but later, parodied and ridiculed. How could this happen? It will be recalled that the Linnaean classification of plants and vegetables with its sexualisation of same, naturally struck a chord in his mind. Here was an opportunity to participate in a sexual, albeit literary *pas de deux* whilst ensuring that all he had in mind, (of course!) was Mother Nature's plant kingdom. His poem was called *The Loves of the Plants*, and was composed concurrently with his work on the Linnaean system. One can imagine the gradual awareness of the variety of plant sexuality inspiring Erasmus' fertile imagination. However, it was to be five years before he submitted it for publication. He was afraid of any adverse effect it might have on his medical career. Why did he not consider a *nom de plume*? Such was apparently not invented until 1823.

Meanwhile, over in Lichfield, Anna Seward had been busy composing a poetic novel, called *Louisa, a Poetic Novel.* Consequently, having achieved national literary fame, she became known as the *Swan of Lichfield.* Darwin then turned his attention to the cotton industry and spinning machinery but hastily retracted his views when it became clear that big money was involved in regard to patents with names such as Arkwright (Cromford), Strutt (Derby) and Need (Nottingham). He cleverly incorporated the mechanism of spinning machinery in *The Loves of the Plants:*

> With quicken'd pace successive rollers move,
> And these retain, and those extend the rove;
> Then fly the spoles, the rapid axles glow,
> And slowly circumvolves the labouring wheel below

(The *rove:* A sliver of cotton, wool or other fibre, drawn out and slightly twisted, especially preparatory to spinning. *Spoles:* The small wheel near the spindle in the common spinning-wheel).

Several years later, **George Canning**, Britain's prime minister together with other co-authors would parody Darwin's verse in *The Loves of the Triangles*:

> The spiral grooves in smooth meanders flow,
> Drags the long chain, the polished axles glow,
> While slowly circumvolves the piece of beef below

In 1787, Darwin was elected to the Medical Society of London, but although he could have subsequently fed at the trough of private medicine there, he declined. In 1787, he wrote to his friend Benjamin Franklin in America whom he much admired, and who played a key part in the Philadelphia Convention leading to the Constitution of the United States.

Possibly his greatest contribution to physical science, was in 1788, when he published his classic paper on the formation of clouds. It was published in the Philosophical Transactions of the Royal Society and commenced thus:

> Having often resolved in my mind the great degree of cold producible by the well known experiments on evaporation; in which, by the expansion of a few drops of ether into vapour, a thermometer may be sunk much below the freezing point; and recollecting at the same time the great quantity of heat which is necessary to evaporate or convert into steam a few ounces of boiling water; I was led to suspect that elastic fluids, when they were mechanically expanded, would attract or absorb heat from the bodies in their vicinity.

He performed several experiments such as firing an air gun into the bulb of a thermometer and detecting a decrease in temperature of 2°F.

In 1789, Darwin's image was threatened by a metamorphosis into the arena of respectable poetry. He realised the risk to his reputation, not just because of roaming into the world of pistils and stamens (which to those of a somewhat prudish nature approximated to pornography as then perceived), but also because it threatened his status and first regard as a doctor. Furthermore, his style of poetry was by rhyming couplets, which would soon be replaced by other styles such as the enjambed couplet, or the triplet. These forms would be used by the so-called Lake poets, i.e. **William Wordsworth**, **Samuel Taylor Coleridge**, and **Robert Southey**. His fame and poetic status came at a price, including that by the aforementioned Canning.

The Loves of the Plants was published anonymously in April 1789. In case anyone had failed to recognise what Linnaeus meant by his system of classification, Erasmus made it crystal clear that plants were divided into 24 classes depending on their sexual vital statistics. Remembering that the male organ was the stamen, one per flower was classified as *Monandria*. Class 2 was *Diandria* etc, class 13 was *Polyandria*. Class 17 was *Diadelphia* (two brotherhoods with many stamens united into two companies), down to *Cryptogamia*, a clandestine state in which the precise numbers could not be readily counted. In his poetry, he mentioned males as 'five hapless swains', or 'four sylvan youths', all in pursuit of the desired female (pistil).

> What Beaux and Beauties crowd the gaudy groves,
> And woo and win their vegetable Loves.

It almost succeeds in making vegetarians of all red-blooded people. Interesting that he believed that plants were able to feel. Naturally, this was supported by the insectivorous plants e.g. the Venus flytrap. He ingeniously was able to link this romancing with science and engineering, viz. the Carline thistle whose plumes drift far in the wind (Mongolfier's balloon), flax and cotton (Arkwright's cotton mill), Papyra (ancient writings). Opium addiction is also mentioned:

> She waves her wand again! – fresh horrors seize
> Their stiffening limbs, their vital currents freeze; [cold turkey?]
> By each cold nymph her marble lover lies,
> And iron slumbers seal their glassy eyes. *[Pin-point pupils, a feature of morphine addiction].*

Also mentioned is '*seductive Vitis*' the vine with her alcoholic poison:

'Drink deep,' she carols, as she waves in air
The mantling goblet, 'and forget your care'...
Fell Gout peeps grinning through the flimsy scene,
And bloated Dropsy pants behind unseen.

Erasmus was very aware and fearful of the effects of alcohol. Escape from the cares of the world, accompanied by gout, and heart failure with generalised oedema ('dropsy'). In a note, he suggests that Prometheus stole a fire (fiery spirit), which was the 'vulture gnawing at his liver' i.e. alcoholic cirrhosis. In regard to the plant Adonis, we find polyandria, together with polygynia in a squat, (many males and females in one flower). They move in a procession to the shrine of *'Licentious Hymen'* attended by *'light Joys on twinkling feet,'* while *'exulting Cupids'* pepper them with *'promiscuous arrows.'* The mind boggles! *The Loves of the Plants* delighted its readers including Horace Walpole, son of Sir Robert Walpole, England's first prime minister. Horace Walpole was an author, who wrote about Gothic themes. He said (in regard to *The Loves of the Plants*);

> You will agree with me that the author is a great poet ... I can read this over and over again forever ..., all is the most lovely poetry

Erasmus was also anti-slavery, along with Wedgwood. There were many wealthy owners of West Indian sugar estates. Unfortunately the issue would not be resolved for 20 years on account of the strong House of Commons lobby, **William Wilberforce** and **Thomas Clarkson** being the main campaigners of the Society for the Abolition of the Slave Trade.

Literary reviews were full of praise, publications such as *Critical Review, Monthly Review, New Annual Register.*

William Cowper was an English poet and hymnodist. One of the most popular poets of his time, Cowper changed the direction of 18th-century nature poetry by writing of everyday life and scenes of the English countryside. He wrote an eight page review, stating that *The Loves of the Plants* was:

> of a very superior cast ... with so much versatility of genius

William Hayley, English poet, biographer, and patron of the arts, also admired the work. Meanwhile what of Anna Seward?

She wrote a 100 page analysis. In it she admitted that the:

> sexual nature of plants has been proved *and* the female form is always attractive from the poetic pencil of Darwin.

One might query her choice of term. Could the *poetic pencil* of Darwin have equally been phrased as '*the poetry of Darwin*'? Or did Darwin's '*poetic pencil*' imply some phallic fascination?

In 1789, Elizabeth bore Erasmus a sixth child, Henry. Although he thrived following birth, sadly he died aged 12 months. The same year, Erasmus' sister Susannah died, aged 60.

Meanwhile, at Full Street, their home in Derby, Erasmus purchased an orchard on the other side of the river Derwent. Access proved no problem, for he designed and produced a self-propelled ferry, using a system of levers and pulleys.

Darwin's restless mind now ventured into the world of geology and chemistry. His interest included the theories of the formation of rocks. In addition he was interested in fossils and chemistry. He composed a 20 page essay on the waters of Buxton and Matlock, published in 1789 as part of an article by James Pilkington titled *View of the Present State of Derbyshire.* Some felt that the warmth of hot, or rather, warm springs, arose from a chemical action, whereas Darwin thought it emanated from heat much deeper down in the Earth's mantle.

Darwin also threw down the gauntlet on the thorny subject of the **phlogiston theory**. The big boys (Priestley, Keir and Watt) held the faith (eventually shown to be mistaken) of this theory. It is a superseded scientific theory which postulated that a fire-like element called *PHLOGISTON* is contained within combustible bodies and released during combustion. Phlogiston was a theory of burning matter that said that it was released from burning material, and was what actually burned. However the new theory said that something, (now known to be oxygen), came out of the air and was added to a material while it was burning. Darwin defected to the **Lavoisier** (French) school. Lavoisier is most noted for his discovery of the role oxygen plays in combustion. He recognised and named it (1778) and hydrogen (1783) and opposed the phlogiston theory. Darwin's *Loves of the Plants* was followed by *The Economy of Vegetation*. In it, the word oxygen proudly reared its head, whereas a few years earlier, it would have invoked ridicule. He also introduced the word hydrogen.

Darwin hoped to include engravings of the **Portland Vase** in this latest work. The Portland Vase is a Roman cameo glass vase, which is dated to between AD 1 and AD 25, though low BC dates have some scholarly support. It is the best known piece of Roman cameo glass and has served as an inspiration to many glass and porcelain makers from about the beginning of the 18th century onwards. However, his publisher **Joseph Johnson** who was an influential 18th-century London bookseller and publisher whose publications covered a wide variety of genres and a broad spectrum of opinions on important issues informed Darwin that there would be legal difficulties if they were copied. However Johnson offered to have new engravings made by **William Blake** who at an early age, had started engraving copies of drawings by the great Old Masters of the Renaissance like Raphael, Michelangelo and Albrecht Durer.

The Economy of Vegetation appeared in 1791 and subsequently was bound to the first edition of *The Loves of the Plants*, (1789) the pair together constituting the *Botanic Garden*. In 1803, (posthumously) appeared part three, *The Temple of Nature*, (the title preferred by Johnson the publisher), though Darwin had intended it to be *The Origin of Society,* the basis of which was Darwin's epiphany of evolution as originating:

> '***from a single living filament, a speck in primeval seas, through fishes and amphibians to land animals and eventuality to humankind,***'

as he named us. It established Darwin as a serious poet. This sequence of three epic poems was edited meticulously by **Stuart Harris**, and has been published under the title *Cosmologia*, the first edition appearing in 2002, and the second in 2004. *The Economy of Vegetation*, in dramatic form describes the birth and evolution of the universe utilising the four elements, fire, earth, water, and air, as originally described by **Empédocles** (born 490 BC) who was a Greek pre-Socratic philosopher and a citizen of Akragas, a Greek city in Sicily. Erasmus no doubt had in mind his friends, Priestley, Wedgwood, Brindley et al and the emerging technology of the time.

A preamble to Cosmologia, actually composed by Stuart Harris goes as follows:

> The doctor's green-gold phaeton, whose spoked wheels's
> Serpentine-springs glide easily over
> The unmade earth of country roads, carries
> Sweetmeats, surgical instruments, mercury;
>
> And books as guidance, for a cosmic journey;
> Bryant's Mythology, Linnaeus, Herschel's
> Star-map, and Newton's Principia;
>
> And words (stuttered maybe): enquiries,
> *Diagnoses offered with sympathy;*
> *Witty exchanges, sarcasm occasionally*
> *Too strong for the listener; passionate appeals*
> *To altruism and our better nature.*

(Here I would like to interpose Richard Dawkins' view of **altruism**, for which he proposed the term 'meme' and the selfish gene, regarding the gene as the *replicator* and the life form merely as the *vehicle* subservient to the gene. What might Erasmus have made of Dawkins' thinking?)

And poetry: artesian well-water
Sustaining a muse who is the goddess
Of botany; the genius loci
Of a star-flower garden whose beauty fills
Every couplet. Wheels mired in mud, the car
Halts by a garden-gate, and the doctor
Strides, with offered apple-tart, towards a lass
Thinned to the bone by work and poverty.

This suggests either that the good doctor is visiting a poor i.e. not paying patient, or else is visiting a servant of one of the well-heeled class.

The Loves of the Plants incorporated a stuttering start by using Anna Seward's 34 lines on the subject of the botanic garden at Lichfield. Perhaps Erasmus intended this as a flattering acknowledgement of her influence, but whatever, she was not best pleased, regarding it as plagiarism. Nevertheless, she later changed her mind, and persuaded herself that it was actually a compliment to her. It utilized the Linnaean taxonomy and dramatized it, then continued with poetic portrayal of amorous-driven emotions, probably variously interpreted by its readers.

Darwin's poetry would unfortunately be eclipsed by Coleridge, and Wordsworth both of whom after initially expressing their esteem subsequently retracted to a degree. They regarded his style as being too dazzling. By contrast Anna Seward stated that:

> not one great poet of England is more original than Darwin

and by including science, his poetry was so brilliant that it would probably never be equalled. It is thanks to the combined efforts of Desmond King-Hele, and Stuart Harris, that the penumbra of his writings has been revealed 200 years after the eclipse. Of course it must be remembered that the reader of today, so inured by previous exposure to the media, may find difficulty in appreciating the subtleties Darwin intended. It is as though pop music that dazzled in the seventies no longer held sway half a century later.

Darwin pondered on Earth sciences. His musings centred on vegetation, geology, and the atmosphere. He stated that the sugar created by vegetables was the energy giving source to the animals that ate them. But he was off the mark in believing that the sugar (i.e. carbohydrate) was produced by decomposition of water. He also believed that anthers and stigma, the male and female structures that by contact led to the production of the seeds within vegetables, were possessed '*with the passion and power of reproduction.*' (But how do we know they weren't? In any event, how would such power be measured, or detected?) Turning to geology, there were two schools of thought, namely the *Neptunists* who stressed the importance of the strata found below the sea, and the *Vulcanists* who laid more emphasis on volcanic events. He was convinced of the high temperature of the Earth's core which included iron which he believed was the source of the Earth's magnetic field (he was right). Next, his interest turned to oil, having been informed of the fountain of fossil tar, or petroleum near a coal seam at Coalbrookdale, near Ironbridge in Shropshire. Tar still oozes through this bitumen spring accidentally struck by 18th-century miners. Up to 480 gallons a day were collected. His attentions also focussed on the composition of the atmosphere, and the formation of clouds.

But science was not the only sphere of life that attracted Darwin. There can be little doubt that the expansion and success of the British Empire was at least in part due to the utilisation of slave labour, and not only by imported black slaves, but also our indigenous white population for whom food and lodging demanded soul destroying work. Even children were employed, and denied education. But Erasmus Darwin, to his credit, was a founder member of the **Derby Society for Political Information**, which was formed in December 1791. In November 1792 Dr W. Johnson a friend of Darwin presented an address to the French National Assembly, and in May 1793, to the House of Commons, who rejected it. Darwin was in danger, and some members of the judiciary sought to make martyrs of those in favour. Transportation to Botany Bay, and likely death, was the punishment. Darwin was targeted as a consequence of the Derby address which appeared as an advertisement in the *Morning Chronicle* on Christmas

Day 1792. A year later, it came to trial as sedition, but eventually reached a verdict of 'not guilty.' Darwin was lucky, because rumour had it that the address had come '*from the pen of a writer, whose productions justly entitle him to rank as the first poet of the age; …*' A flattering threat. Given the climate of repression pervading in 1793-1794, had his letters been publicised, he would have been forcibly embarked on a ship bound for Australia.

Poet, geologist, balloon expert, meteorologist, fertile father, let us not forget that first and foremost he was a doctor. He was hard at work on *Zoonomia*, volume 1 in 1794, and volume 2 in 1796. These included his theories on medicine and life. It is an awesome thing to imagine how anyone could command the respect he did when so little was known and of proven value in contrast with today. People including babies and children just died, medical attention (however lacking) was mainly the privilege of the rich. The poor had to do without, and their protests, if any, would be unlikely to be heard. In the case of the wealthy and influential patient, more effort was required, and was forthcoming, though not necessarily of any use. If a cure became apparent, it was more likely the result of Mother Nature, than of the medication prescribed. One development that was discovered, though of limited or no value at the time, was pneumatic medicine. **Dr Thomas Beddoes** considered that the breathing of certain air mixtures might be beneficial. It was tried as a form of therapy for tuberculosis but did not prove of value. Darwin felt that pupils needed to breathe pure air at school and in their bedrooms. This was probably very appropriate in the vicinity of the Wedgwood's pottery industry in Etruria. Nowadays, air pollution is everywhere in our towns and cities. Stoke was 200 years ahead of most of the world on this topic, but not in a good way. He also had views on exercise, being of the opinion that most pupils were too sedentary. He recommended ball games, and the use of swings. Intriguingly he gave his views on care of body shape, especially girls at boarding schools. He was in favour of doing away with stays, and instead they should appear naturally graceful. He was ahead of his time, and probably appreciated the healthy female body form bereft of stays and obesity. He also expressed views on nail biting and twitching. Fidgeting also, though we now know that the latter could have been a symptom of acute rheumatic fever, (St Vitus' dance) about which little was then known. Meanwhile, Jessie Watt, wife of James, of steam engine fame, had died of consumption (TB). Following his wife's death, Watt joined the quest for production of pure gases that might be of benefit in disease. Beddoes proposed a Pneumatic Institution in Derby and Thomas Wedgwood gifted the sum of £1,000 pounds. Darwin then exposed a weakness in his medical treatment. For so many symptoms, his prescription was opium. Of course, this was likely to make patients feel better, at least at first. Often increasing doses were required, and addiction was the result. (Nowadays, a similar sequence of events may follow the ill-judged over-prescribing of steroids though falling short of actual addiction). However, what choice did he have? The medical formulary was like Old Mother Hubbard's cupboard, consisting largely of emetics, purgatives, and opioids. Incidentally, it also included digitalis but only one person really understood its dangers and potential for success, and that was not Darwin. It was Withering, a fact that will be revealed in greater detail in the next section on Robert Darwin.

Erasmus Darwin's *Zoonomia*, also called *The Laws of Organic Life* appeared in the first half of 1794. It consisted of an attempt to classify animals in much the same way as Linnaeus had earlier classified plants. He commenced by explaining bodily action ('fibrous motions') into four classes;

> Irritation, sensation, volition, and association.

In his way, he was classifying the body's responses to be a consequence of *irritation* e.g. pain production by external influences (nowadays termed 'nociceptive'), *sensation,* (pleasure or pain), *volition* (positive by desire, or negative by aversion). He regarded drunkenness as a disease of irritation that could result in mental change resembling idiocy, and temporary weakness of the limbs, resembling a stroke. (In contemporary terms, a person may be described as 'paralytic' having consumed alcohol to the extent of not being able to stand up, let alone walk). He went on (correctly) to predict permanent damage by continuous exposure to excess of alcohol. *Association* refers to symptoms that are associated with the primary condition, e.g. vomiting, or fever. Volume 1 of *Zoonomia* comprised 40 essays which ranged on biological subjects such as instinct, sleep, reverie, vertigo, circulation, saliva, stomach, liver, temperament and digestion. Sleep and reverie morph into the field of psychology. On a more obvious note is his chapter on oxygen and blood:

> It appears that the basis of atmospherical air, called oxygene, is received by the blood through the membranes of the lungs; and that by this addition the colour of the blood is changed from a dark to a light red.

He goes on quite correctly to apply the same theory to gills of fish, and the placenta.

Darwin, no doubt through the grape-vine, or his readings, or whatever, got to hear about electricity, and this sparked (pun intended!) his prediction of its future importance. (**Alessandro Volta** was a physicist, chemist and a pioneer of electrical science. He had invented the electric battery then called the "voltaic pile" – in 1800. Using his invention, scientists were able to produce steady flows of electric current for the first time, unleashing a wave of new discoveries and technologies). He stated that:

> it deserves further investigation, as it may acquaint us with new properties of matter.

His essays were not exempt from criticism. With particular regard to the biology of instinct, he provocatively suggested that:

> Hence at our mature years, when any object of vision is presented … which by its waving or spiral lines bears any resemblance to the form of the female bosom … we feel a general glow of delight which seems to influence all our senses … This animal attraction

> is love … the purest source of human felicity, the cordial drop in the otherwise vapid cup of life.

Upon reflection, and prompted by this statement, could it be that our appreciation of form, ('waving or spiral lines'), touch, warmth, and taste might originally have come from being fed at the breast?

Darwin appears not to have been privy to the role of germs. Dutch scientist Antonie van Leeuwenhoek designed high-powered single lens microscopes in the 1670s. With these he was the first to describe sperm (or spermatozoa) from dogs and humans. He also studied yeast, red blood cells, bacteria from the mouth and protozoa. So why didn't Darwin use a microscope to study disease? Did he possess one? Would it have been technically capable of seeing bacteria? Bacteria are universally visualized by the application of the Gram stain, so called after its inventor Hans Christian Gram, who produced it in 1882. Bacteria generally stain purple-brown (Gram positive), or red (Gram negative). An example of the former

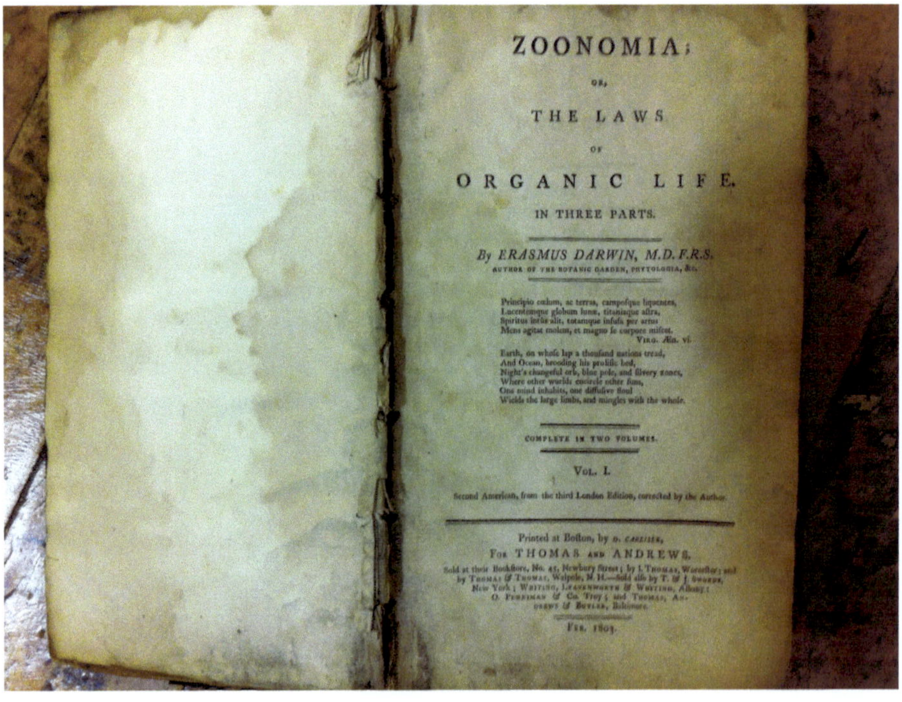

An 1803 edition, (Volume 1) as offered for sale in 2018

is the streptococcus which is notorious for causing severe bacterial sore throat, and of the latter, urinary tract infection by *Escherichia coli*.

The book was (at that time), regarded as an unqualified success by the *European Magazine* which stated:

> [Zoonomia] bids fair to do for Medicine what Sir Isaac Newton's Principia has done for Natural Philosophy.

All was looking well, except for one thing – the French Revolution of 1789, and its protracted after effects, in 1794. In November 1793, the '*British Convention of the Delegates of the People, associated to obtain Universal Suffrage and Annual Parliaments*' was held in Edinburgh. This was regarded as sedition, and would incur deportation for the guilty. Joseph Priestley emigrated to America, not only for this, but also because his house in Fairhill (nowadays Sparkbrook), had been burned down by a mob. Darwin and his medical colleague, Thomas Beddoes were the object of antagonistic poetry in Darwin's style, purportedly written by Darwin, but clearly not so. Britain was in the grip of a repressive atmosphere, tempting Darwin himself to contemplate emigrating to America.

A fad that was becoming popular at that time, was the use of inhaled oxygen. It was used for a variety of conditions, including headache. Darwin treated his own daughter's 'nervous cough' with it, plus opium!

Cataracts then, as now, were very common, and unlike now, there was no treatment for the opaque lens. Darwin, with brilliant foresight said:

> Could not a bit of glass ground into a shape like a shirt-sleeve stud be put into a hole made in the centre of the cornea, and left there, which would stay, and become a glass eye?

Nowadays, of course, we have lens implants, which are hugely successful, to which this author can bear testament!

In 1796, Darwin in a letter to his son Robert stated that he had been attending Robert's grandmother in Sleaford, and:

> who is recovered, for the present at least, from an anasarca (dropsy), I suppose, of her lungs and limbs.

This suggests that she was being treated by someone else, not Erasmus. One is curious whether this treatment by any chance involved the administration of foxglove leaf (digitalis), with its awkward memories of conflict with Withering. The same year, Joseph Wright was treated for dropsy by Darwin, using foxglove. It cured his orthopnoea (the medical term for inability to breathe lying down, necessitating sitting up). This is a feature of heart failure along with generalised oedema (fluid in the tissues), known then as dropsy.

Erasmus Darwin had a mind bent on classifying (diseases), as well as writing, meaning giving a numeric value to paragraphs, sections, and subsections. The former was a heroic attempt at a time when so little was actually known about pathology, and nothing about bacteriology and virology. So everything was based on description, which meant relying on observation of the physical appearance of the patient and his or her actions, as well as their habits and behaviour. He (rightly in this author's opinion), blames credulity to be a deplorable habit, when applied to religion and politics.

At around that time, James Watt designed a centrifuge for whirling patients around in what he termed a 'rotative couch'. Some no doubt felt giddy as a result, and possibly felt the experience called 'vertigo'. A similar event as that experienced by children riding on a rotating roundabout in a park.

Darwin highlighted vertigo as a *natural expression of life* and regarded it as a physiological, i.e. normal phenomenon. Serendipitously, it was during a sea voyage that Darwin's attention was drawn to dizziness. In the chapter *Of Vertigo* in the *Zoonomia* under the caption *Attention of the Mind Prevents Slight Sea-Sickness,* he wrote:

> In an open boat passing from Leath to Kinghorn in Scotland, a sudden change of wind shook the undistended sail, and stopt (sic)

> our boat; from this unusual movement the passengers all vomited except myself. I observed, that the undulation of the ship, and the instability of all visible objects, inclined me strongly to be sick; and this continued or increased, when I closed my eyes, but as often as I bent my attention with energy on the management and mechanism of the ropes and sails, the sickness ceased; and recurred again, as often as I relaxed this attention.

Darwin found that fixing the gaze at one point was an effective means against seasickness, and against visual vertigo in the first place. The anecdotal account was all the more important as it contained the explanatory model Darwin would later apply to the causes of vertigo.

Crucially these would tend to occur as a consequence of visual perception and, subsequently dizziness in unfamiliar movement and motion sequences, i.e. in a lack of routine, practice, habit, or experience.

Darwin continued:

> If first a European mounts an elephant sixteen feet high, and whose mode of motion he is not accustomed to, the objects seem to undulate, as he passes, and he frequently becomes sick and vertiginous. Any other unusual movement of our bodies has the same effect, as riding backwards in a coach, swinging on a rope, turning round swiftly on one leg, seating on the ice, and a thousand others.

Accordingly he recommended to prevent sea sickness it was probable, that the habit of swinging for a week or two before going on shipboard might be of service.

His observations exhibited a remarkable degree of perception and formed part of a scholarly article on rotation and vertigo, quoted in a paper

> The roots of vestibular and oculomotor research,

which appeared in a journal titled *Human Neurobiology*, 1984, volume 3, pages 121-128, author B. Cohen of the Mount Sinai School of Medicine, New York. Praise indeed for Dr Darwin!

Insofar as restoring health to people, in volume 2 of *Zoonomia*, he rated nutrition highly. He noted that populations remained small when food was only obtained by hunting, but farming vegetables and animals conferred greater benefit. He realised the importance of good soil quality as well as that of the animals that grazed on it. He continued:

> To secure the greatest sum of human happiness, there should be no slavery … and no despotism.

He admitted that inequality in the ranks of society was necessary for success, but that this inequality should not be too great. (Clearly he did not subscribe to the belief that all men are created equal. At least he was honest!). Under the heading nutrition he included blood transfusions, and oxygen. Plus opium, alcohol, love and joy!

1797 was punctuated by the deaths of three significant friends, namely Sir Robert Burdett, Horace Walpole, and James Hutton. In addition, his mother died aged 94 at Sleaford, having produced seven children all of whom had thrived. Darwin was busy writing his book on plant life, *Phytologia*. The same year, two Derbyshire cotton mill founder fathers, Arkwright and Strutt died. King George III had stated:

> Why does not Dr Darwin come to London? He shall be my physician if he comes.

It would have been a most challenging task, as the king's illness was poorly understood then and still is to this day, though porphyria has been mooted. His actual physician was the celebrated **Dr Richard Warren**, whose status resulted in immense wealth.

The below paragraph is unashamedly plagiarised from Desmond King-Hele's book on Darwin:

"In June 1797 Darwin received a visit from an unknown gentleman who had just arrived in Derby. He said:

> 'I am come from London to consult you, as the greatest physician in the world, to hear from you if there is any hope in my case. It is of the utmost importance for me to settle my worldly affairs immediately; therefore I trust that you will not deceive me, but tell without hesitation your candid opinion.'

After examining him, Darwin had to say there was no hope; he could not expect to live more than a fortnight. The man seized Darwin's hand and said:

> 'Thank you, doctor, I thank you; my mind is satisfied; I now know there is no hope for me.'

Darwin asked him:

> 'But as you come from London, why did you not consult Dr Warren, so celebrated a physician?'

He replied:

> 'Alas! Doctor, I am Dr Warren.'

He died a week or two later."

One wonders by how much anyone's life might have been shortened in 1797 by travelling by coach to Derby from London and back!

Fame and success can be the harbinger of more of the same, or the opposite. This was to be Darwin's experience. First, his reputation as a doctor, the author of *Zoonomia*, attracted the attention of a precocious young man, who fancied himself as a literary critic, initially a student of law, and then of medicine. His name was **Thomas Brown**.

The tactic employed when a more penetrating attack is intended, consists of an apparently complimentary letter, but then the sword is gradually extracted from the scabbard, and plunged into the victim. He commenced by stating that having read *Zoonomia* with great pleasure, he nevertheless felt that Darwin's reasoning was rather fanciful, and he felt it incumbent to inform the press of his opinion. Brown sent a manuscript of no less than about a thousand pages to Darwin, each page logical and reasoned. Darwin realised he had met his match. In May 1798, *Brown's Observations on the Zoonomia of Erasmus Darwin* was published, and ran to 560 pages. Whatever Brown hoped to gain, and however brilliant his massacring of Darwin's work was, it had the opposite effect, by publicising it. However, more was to come, this time an attack on Darwin's poetry.

George Canning, Under Secretary for Foreign Affairs in the government of **William Pitt**, and who would one day be Prime Minister, saw a threat to established religion by Darwin's presentation of evolution. It must be admitted that Canning, assisted by some collaborators, produced a brilliant parody on *The Loves of the Plants*, which they called *The Loves of the Triangles*. To make matters worse, it was written in Darwinian verse. Instead of nymphs and swains, we had a *parabola*, a *hyperbola*, and an *ellipse* all desperate for the love of a *rectangle*! Brilliant! By way of explanation for those determined to see the serious side, Canning ridiculed Darwin about evolution from lower forms of life, that electricity would ever assume any importance, and that the mountains of Earth were more than the 6,000 years old as gleaned from a reading of the Bible. Eventually of course, Canning was proved wrong and was himself the object of ridicule. Two attacks gone, one was to follow. Darwin's book on medicine, was the first, then his classification and sexualisation of plants, then came the attack on his poetry.

Whereas Wordsworth and Coleridge earlier had been impressed by Darwin's style, with more experience of the world of poetry, so were his perceived weaknesses revealed, and horror! – Lampooned! This was in a little book by both of them titled *Lyrical Ballads*. Henceforth poetry Darwin style would only appeal to older people, the new mode was being established. Yet his influence on both the young bucks was still apparent in

The Rime of the Ancient Mariner (Coleridge), and *Lines written a few miles above Tintern Abbey* (Wordsworth). Darwin's poems feature underground rivers, domes and caverns from experiences in Derbyshire, whilst Coleridge in *Kubla Khan* breathed forth with:

In Xanadu did Kubla Khan
A stately pleasure-dome decree:
Where Alph, the sacred river, ran
Through caverns measureless to man
Down to a sunless sea.

Come 1798, and Darwin's literary output continued with an American edition of *The Botanic Garden*; *Female Education* published in Ireland, and in America, a third edition of *Zoonomia*, and *Phytologia*.

An American admirer, **Samuel Mitchill**, wrote a poem in Darwinian style about *Septon*, the essence of septicaemia, and the enemy of oxygen. This is significant in this day and age in the field of medicine where the fear of sepsis, and the body's defensive collapse due to being overwhelmed, is much to the fore.

Actual, as apart from metaphorical, battles were also being fought on land, the **battle of Killala** which involved Irish rebels aiding the French who invaded at Killala Bay. They were brutally repulsed by the English.

In 1799, an even worse enemy reared its ugly head – income tax! This was introduced during the government under William Pitt. Erasmus wrote to his son Robert, physician in Shrewsbury whose life will be described in the next chapter, that he reckoned he had earned about £1,000, deducting expenses of £400 for travel expenses. Curiously, and not unexpectedly, the income commissioners made a fuss, but eventually calculated much the same. What's new?!

At about this time, a new treatment, or better named, prevention, of smallpox became recognised, although treatment by **variolation** had been practised for centuries originally by the Chinese. Variolation involved rubbing dried scabs from smallpox cases into the skin. It could, and did prove dangerous, but many people believed it to be worth the risk of avoiding a fatal, or if not fatal, disfiguring disease. However, the most significant discovery was by a Gloucestershire GP, **Edward Jenner**, who was told by a milkmaid that she didn't want to be variolated because it was common knowledge amongst milkmaids that they could not get the smallpox because they had previously been infected with the cowpox, (a mild complaint) caught from cows' udders. Jenner thought long and hard, and corresponded with the esteemed anatomist John Hunter, who suggested he proceed using cowpox material, which he did, and the rest is history. Simon Phipps was vaccinated using scabs from cowpox, and later challenged by variolation. He didn't subsequently exhibit any signs of illness attributable to smallpox. This process was given the name **vaccination**, from the Latin *vacca*, meaning cow. There were however many who objected, particularly those who made a living out of variolation. And in today's world, no ethics committee would have granted permission for such an experiment. And did Simon give his informed consent?

As noted earlier, when Erasmus' son Charles, whom Erasmus recognised as the brightest of his offspring, was studying medicine at Edinburgh University, he was given the opportunity to carry out a post-mortem examination as part of his training. However, tragically, Charles developed a fatal infection from the corpse. Unfortunately, knowledge about bacterial infection, and its prevention, had not yet been acquired by the medical profession – it would be another century before **Joseph Lister's** teachings regarding anti-sepsis would reach the world.

Another son, Erasmus, the lawyer, having reached the ripe age of 40, decided to retire. He let it be known that he wished to build a cottage for this purpose. He purchased a farm near Derby, called the Priory. Erasmus Senior was none too pleased, as he believed in hard work, in many avenues of life. Erasmus Junior was clearly fed up with the drudge of legal documents. He had allowed a large amount of his work to accumulate, but had given his word that he would complete sending out bills before the end of the year. To this end, he had worked continuously for two nights, refusing rest and food. The succeeding events are open to conjecture, but the upshot was that his body was found in the River Derwent. Naturally his father was destroyed by the news, and intriguingly Anna Seward accused him of having starved his son of affection, possibly leading to the tragedy. This smacks of malice, perhaps incubated over the years, and dating back to a time when their names (Erasmus Senior) and Anna might have been linked. The actual cause of death by drowning was either suicide or misadventure. And Erasmus himself did not know he had but two years left in this life.

Not only was he left with the memories of his sons, but also the unfinished accounts. They stretched back over 10 years! And to make matters worse, there were several outstanding debts. The Priory recently purchased by Erasmus Junior was willed to his father, who chose to retire to it with Elizabeth. He made several alterations to it. His wife wished to escape the morbid atmosphere, so took her daughters to stay with Robert Darwin and his wife Susannah in Shrewsbury. Then on to Barmouth for a fortnight's holiday, before returning to Derby, followed by Buxton, Chatsworth and Matlock. One is left wondering whether Erasmus might not have appreciated his wife's support at this sad time.

Darwin's most important non-poetical book *Phytologia; or the Philosophy of Agriculture and Gardening* was published in 1800. It dealt with photosynthesis and plant nutrition. The first part was concerned with physiology of vegetation, i.e. structure and function. He voiced some original views such as that:

> *the cold and motionless fibres of plants*

were also alive, and:

> *vegetables are in reality an inferior order of animals.*

Weird! Whilst having nothing against the practice of taking buds to grow plants, he was clearly cognisant of sexual reproduction of plants, courtesy of Linnaeus,

Sexual reproduction is the chef d'oeuvre, the master-piece of nature.

He was aware that leaves are the organs of plant respiration having, like the lungs of animals, a huge surface area, and he hypothesised that they had minute pores. To test this hypothesis, he covered some leaves with oil in the belief that this would block the pores, inhibit respiration, and lead to the death of the leaf. He was right. Thirty years later, this was confirmed by others, the pores being termed *stomata*. He also thought that sap was part of a system of circulation by arteries and veins resembling that of animals. In this he was to a point also right, although there are no arteries and veins as such, but instead there is **phloem** and **xylem**. (During transpiration plants move water from the roots to their leaves for photosynthesis in xylem vessels. Glucose made in photosynthesis is then moved to all cells in phloem vessels for respiration). He observed first hand that manure helped plants grow, and that carbonic acid and water were essential for growth. This was the source of carbon dioxide. When vegetable leaves were exposed to sunlight they released oxygen. (Dead right, and on this our planet survives, but for how long, with large scale deforestation?)

The essential function and naming of chlorophyll would have to wait 40 years, but meanwhile it was called *green matter*. From observation that putrefaction resulted in the smell of ammonia, he concluded that nitrogen (a known component of ammonia) was involved. He also deduced that phosphorus was important, and he was right, although his thinking was based on the observation of phosphorescence by putrefying wood or meat e.g. fish heads in Edinburgh streets. However phosphorescence is based on chemiluminescence, and phosphorus is not always a necessary requisite for it. Chemiluminescence is the production of light from a chemical reaction. Two chemicals react to form an excited (high-energy) intermediate, which breaks down releasing some of its energy as photons of light. Chemiluminescence is currently used in many laboratory tests. He also realised the future value of sewage farms, but well away from towns, for obvious reasons.

For control of plant diseases caused by aphids, he recommended the introduction of another insect that would feed on aphids i.e. the biological control of insects for which he became rightly recognised as a pioneer.

Darwin ventured into the pastures of fancy, with his concept of organic happiness. Nature, he admitted is savage, vegetables being eaten by animals, and animals (in whom pain is instantly recognisable) being eaten by other animals. Organic happiness he supposed came with the success born of evolution, the higher animals having a greater capacity for it. However he disregarded the corollary that so does misery.

Come March 1801, he became ill possibly with pneumonia and pleurisy. He treated himself, and communicated details to son Robert, the doctor in Shrewsbury. Some details are of interest:

> *As the pulse is so regular, I don't think there is any chance of water in the chest, and as it is so slow I do not think there is any abscess, but that simply it is the continuance of inflammation.*

He bled himself once every three or four days, and took 20 drops twice a day of tincture of foxglove. It is quite likely that with a regular pulse, foxglove (the source of digitalis) would be of little or no benefit (we nowadays know this to be so, as it works best in the presence of an irregularity called *atrial fibrillation*. The pulse being slow might have been a side effect of the digitalis, which can be very toxic). It was his *bete-noire*, William Withering, who knew all about the whys and wherefores of foxglove. In April he wrote that his pulse kept missing a beat, every second, third or fourth. These are also possible side effects of digitalis. Ironic that the doctor who reckoned himself to be on a level with Withering on the subject of use of foxglove leaf, and who had on one occasion carried out a joint consultation on a patient, should himself be the subject of relative inexperience and expertise and consequently the object of a side effect.

In 1801, the third edition of *Zoonomia* appeared. He expanded his concept of evolution viz. the contest of plants for light and air above ground, and food and moisture below. He also recognised the possibility of the extinction of species by noting the abundance of shells and other fossils in limestone, but no longer in the oceans. By imagining the external features, *phenotype,* of the offspring to be a consequence of the conjoining of fibrils or molecules i.e. small particles contributed by each parent, he was unconsciously anticipating the role of chromosomes (and within them, genes), from each parent, and of course the message from each that is carried in the DNA, this nowadays being known as the *genotype.*

In 1802, circa 22nd March, Erasmus and family moved to Breadsall Priory in Full Street, Derby. According to an advertisement in the Derby Mercury, it was:

> *A large excellent dwelling with a Brewhouse and all convenient offices, a five walled stable, lounge, pleasure and kitchen gardens, the house pleasantly set at the lower end of Full Street.*

Unfortunately, his genius as revealed in *Zoonomia*, was ill-timed, as the country was more concerned with the consequences of the French Revolution, impending invasion by the French and the hope that God would protect. Not exactly the best occasion to be given a scientific explanation very different from that of Divine creation. Upon reflection one sees the evolving Darwin, from poet extraordinary to scientist.

However, two weeks later, this is what happened, according to son Robert:

> *On 10th April, he was attacked with a severe cold fit of fever, followed with a proportionate hot fit, and with feelings and symptoms that threatened an inflammation of the lungs … He was bled … 25 ounces … got well in two days. But on 18th April died having partially finished a letter he was writing.*

At the time of his death, Elizabeth was 54, and the mother of seven children by him, and four by her first husband.

Breadsall Priory (House and heritage.org, with permission)

Somewhat surprisingly a post-mortem was carried out. His heart appeared in good condition, there was some adhesion of the pleura to the right lung. He may have died of pneumonia, but equally possible is that he had a gram-negative bacteraemia due to an infection of the urinary tract, causing sepsis.

He was buried at Breadsall Church, His memorial inscription reads:

Of the rare union of Talents
Which so eminently distinguished him
As a Physician, a Poet and Philosopher
His writings remain
A public and unfading testimony
His Widow
Has erected this monument
In memory of
The zealous benevolence of his disposition
The active humanity of his conduct,
And the many private virtues
Which adorned his character.

But the Darwin genes were set for possibly greater things, though only indirectly via the next generation, i.e. Robert, whose story follows.

Chapter 4

Robert Waring Darwin (1766-1848)

Robert Darwin was born in Lichfield and named after his uncle, Robert Waring Darwin (1724-1816). He was the fourth child and third son of Erasmus and his first wife, Mary (Polly) Howard. His mother died in 1770, probably of gallbladder disease. Because she had severe pain requiring large doses of morphine and alcohol, she was suspected of alcoholism. Subsequently **Mary Parker**, the governess hired to look after him, became his father's mistress and bore Erasmus two illegitimate daughters. In 1778, tragedy struck the family. Charles, Robert's elder brother, a brilliant medical student at Edinburgh, died after catching an infection from a post-mortem he was carrying out as part of his training. Erasmus was mortified, as he had great hopes for his firstborn. He then turned his aspirations for success in medicine, to his third son, Robert.

Portrait of Robert Waring Darwin by J.C. Pardon (Courtesy of Shrewsbury Museums service)

In 1782, aged nearly 16, he was sent to attend anatomy classes in London. They were probably given by the famous anatomist **John Hunter (1728-1793)**. Afterwards, he went on to Edinburgh, which was easily the leading British medical school at the time, and where he obtained his medical degree MD in 1785. Although Oxford and Cambridge might have seemed the obvious choice, they were at that time, preoccupied with religion and the classics, rather than cutting edge science. Erasmus realised this and directed his son northwards.

To broaden one's education further, it was felt that one had to visit reputable European centres, and at this time, **Leiden University** was the choice. It owed its claim to fame to **Herman Boerhaave** (1668-1738) who was a Dutch botanist, chemist, Christian humanist, and physician of European fame based there. He was regarded as the founder of clinical teaching and of the modern academic hospital. Also he introduced the quantitative approach into medicine, and is best known for demonstrating the relation of symptoms to lesions. Additionally he was the first to isolate the chemical urea from urine, as well as the first physician to put thermometer measurements into clinical practice. So Robert went to Leiden University in 1784 and obtained the degree Doctor of Medicine, MD, in 1785. The question then arose as to where to start out in medical practice. He got to hear that one of the doctors in Shrewsbury had died, so together with his brother (not his father), Erasmus, he went to have a look. It was to his advantage that his father was an acquaintance of the Clive family who lived in Market Drayton, a few miles from Shrewsbury. They stayed there a while. Robert Clive, son of Lord Clive, was impressed by Robert Darwin, and this was to be a definite advantage in the future setting up of his medical practice. Erasmus, Robert's father, and Josiah Wedgwood each donated £20 to help set up the practice, together with a bay mare and a chestnut horse, so that patients would be impressed when he was making house calls. (This author recalls how impressed he was by his excellent GP, Dr H.W. Bambridge, of Wellington, Shropshire, making a house call in 1951 in a shining black Jaguar Mark 7 saloon. So lasting was the impression, that it no doubt influenced him to become a doctor 12 years later!)

Dr Darwin also became a freemason, along with Thomas Telford, and William Hazledine, millwright and ironmaster. These connections, among others, no doubt helped build his practice, which thrived, acquiring

50 patients within six months. In the eyes of his father, however, merely being a successful doctor was not enough. Erasmus, as has been described, was a great author, poet, inventor as well as successful medical doctor. Therefore, Erasmus tried hard to enable son Robert to be recognised in the scientific world, by contriving to get him elected to become a Fellow of the Royal Society, (FRS).

The Mount, Shrewsbury. The home of Dr Robert Darwin (Courtesy of Shrewsbury Museums service)

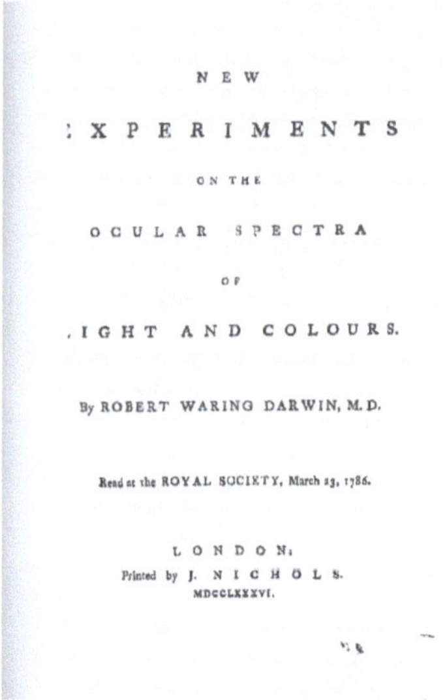

NEW

XPERIMENTS

ON THE

OCULAR SPECTRA

OF

IGHT AND COLOURS.

By ROBERT WARING DARWIN, M.D.

Read at the ROYAL SOCIETY, March 23, 1786.

LONDON:

Printed by J. NICHOLS.

MDCCLXXXVI.

Robert, or was it Erasmus Darwin's submission to the Royal Society?

Look at this for about 30 seconds, than close your eyes, and record what follows after about 15 seconds.

When any one has long and attentively looked at a bright object, as at the setting sun, on closing his eyes, or removing them, an image, which resembles in form the object he was attending to, continues for some time to be visible: this appearance in the eye we shall call the ocular spectra of that object.

But note the date and place from which the paper was despatched to the Royal Society. It suggested that Robert was living at his father's house in Derby, and either alone, or with his advice, composed the work which later that year, November 1785, was submitted to the Royal Society.

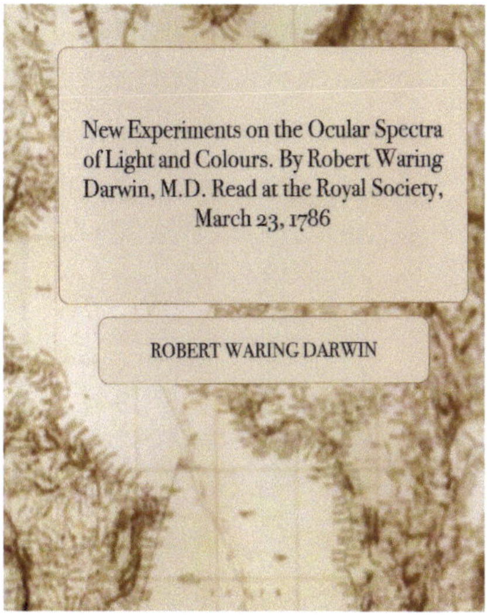

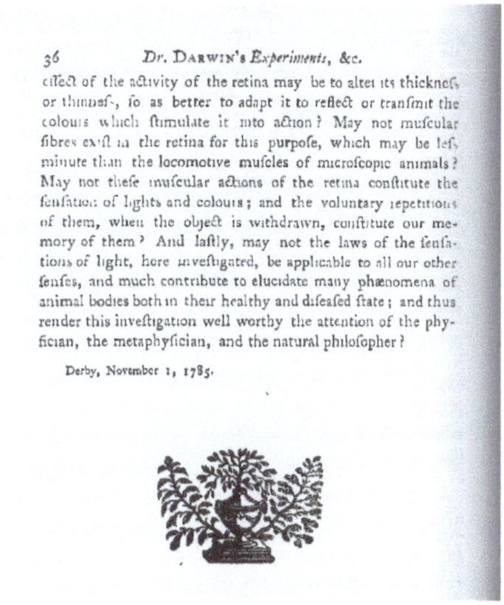

36 Dr. DARWIN's *Experiments*, &c.

effect of the activity of the retina may be to alter its thickness or thinness, so as better to adapt it to reflect or transmit the colours which stimulate it into action? May not muscular fibres exist in the retina for this purpose, which may be less minute than the locomotive muscles of microscopic animals? May not these muscular actions of the retina constitute the sensation of lights and colours; and the voluntary repetitions of them, when the object is withdrawn, constitute our memory of them? And lastly, may not the laws of the sensations of light, here investigated, be applicable to all our other senses, and much contribute to elucidate many phænomena of animal bodies both in their healthy and diseased state; and thus render this investigation well worthy the attention of the physician, the metaphysician, and the natural philosopher?

Derby, November 1, 1785.

Robert Darwin provided the first empirical evidence that small eye movements occur even when people attempt to keep them fixed. This he found during his studies of the after-images of coloured stimuli in which he noticed that while a person tried to fixate a red circle, a lucid edge appeared on the adjacent white-paper background. He concluded:

> *as by the unsteadiness of the eye a part of the fatigued retina falls on the white paper.*

Here is what Charles, Robert's famous son wrote about his father:

> *The third son, Robert Waring Darwin, (my father born 1766), did not inherit from his father any aptitude for poetry or mechanics, nor did he possess, as I think, a scientific mind, though he had a strong taste for flowers and gardening.*

What a crushing comment! It could have been countered by at least a comment on his qualities as a medical doctor, but none was forthcoming. Maybe Charles didn't have a particularly high regard for the medical profession which at that time didn't make much use of science.

> He published ... a paper on Ocular Spectra ... which I believe was largely aided in writing it by his father. He was elected a Fellow of the Royal Society in 1788.

This was two years after the paper had been read! Was this the usual practice, or is it possible that there was some query about its true authorship? The prestigious award of Fellowship of the Royal Society (FRS) is today regarded as the pinnacle of achievement, only slightly trumped by the Nobel Prize. But did Robert Darwin actually deserve it? The title of the paper is shown in the figure, but it does not say who actually read it to the society.

In October 1788, he was appointed physician to the Salop Infirmary, later the Royal Salop Infirmary. On 18th April 1796 he married **Susannah Wedgwood** daughter of the pottery manufacturer Josiah Wedgwood and eventually they had six children. As a child, he had been educated with the young Wedgwoods at Etruria Hall the Wedgwood's family home. Mary Parker, the governess hired to look after him, became his father's mistress and bore Erasmus two illegitimate daughters. At age 15 years, his father married a widow, **Elizabeth Pole** 16 years his junior. This seems not to have had any serious repercussions among his offspring.

It is a curious fact that Robert wasn't keen to practise medicine at least in the early days. His son, Charles wrote that he disliked his profession, but clearly grew to accept it, as his numerous patients accepted him. He became wealthy (which might have played a part). And with wealth came pride.

But while Robert was working hard at his medical practice, he was also the father of an expanding family. He and his wife Susannah were destined to have seven children. She had been seriously ill during her third and fourth pregnancies, and here they were five years later when once more she was expecting. However, their fears were ill-grounded, and Charles Robert was born on Sunday 12th February 1809. It will be recalled that the Charles Darwin of the previous generation, Robert's brother, had died in Edinburgh whilst studying medicine, indeed he had been the apple of Erasmus' eye, and it was clear that Erasmus didn't realistically expect much from Robert, over and above practicing medicine. And so it would turn out, except that he would father the most famous Darwin of all, who unlike father and grandfather, would not study or practise medicine.

For any new doctor of the time, patients meant money. Doctors were unsalaried, and needed to establish suitable contacts. This naturally didn't include the poor, although, once they were financially secure, they could afford to offer free treatment, and thus further enhance their reputation. Erasmus knew well how to achieve local fame, and recommended 'visibility and sociability'. Getting around meant mixing with well-to-do farmers. Remember, Shrewsbury was a thriving market town. But there was to be a cloud on the horizon, and events that would follow were described and analysed by Posner, in 1975 as 'The battle of Wellington' in a paper read to the British Society for the History of Science.

The young Dr Robert Darwin, aged 22yrs, was called to see a Mrs Houlston, a well-known publisher and bookseller in Wellington, a market town about 12 miles away. (And, incidentally, the town in which this author lived from 1947 to 1958). The precise nature of her complaint was, and still is, unknown. As in 2019, patients nowadays tend to visit their chemists for simple remedies, thus it was then. But the apothecary of the time, was worried about the progress of the patient, so called Dr Darwin to visit. This he did, and prescribed according to his judgement. No actual diagnosis was pronounced, a fact that was not unusual in those times when investigations weren't thought of, bacteria hadn't been visualised, there were no antibiotics, but malaria fevers were treated with Peruvian bark, and later, other fevers were successfully treated with willow bark from which aspirin would eventually be discovered. The apothecary however felt that Mrs Houlston's condition was giving increasing grounds for concern. What to do? He remembered that Edmund Withering, another local apothecary who had died in 1769, had a son, William Withering, graduated from Edinburgh, and currently a leading doctor based in Birmingham. So he asked Dr Withering to see Mrs Houlston. Whether or not he happened to mention that she had recently been seen by Dr Darwin we don't know. Anyway, Withering travelled the 30 or so miles from Birmingham, saw the patient, and changed her treatment. She eventually got better. Withering's mistake was in not politely informing Robert Darwin of his opinion, and in not discussing her management. But that was not the end of it. Robert Darwin, probably aided and abetted by his father, was furious, and there followed an exchange of acrimonious letters. The Darwin-Withering axis is an interesting one. It will be remembered that Erasmus Darwin, after the death of a leading Birmingham physician, William Small, had invited Withering, at the time a physician newly graduated and working in Stafford, to apply for the William Small vacancy. Thus Withering was appointed to this prestigious post. The feud between the two led to an awkward situation within the Lunar Society which regularly met in Birmingham. The erstwhile atmosphere of scientific interest and bonhomie at these meetings became awkward. In October 1788, Robert wrote thus to Withering:

> *Your ungenteel medical behaviour to me in the case of Mrs. Houlston in not acquainting me that you was there, when you knew I was in the town and had previously concerned with her, was not very consistent with the character of a gentleman, but the activity with which you have slandered me on this occasion is a matter of much more serious consequence. I understand you told Mrs. Houlston that she had no fever when her pulse was 126 in a minute; and that she had an inflammation in her liver when she had neither pain, soreness or fullness in the hepatic region, nor any yellowness of her skin, or any increased evacuation!*

Darwin thought she had typhus (a condition transmitted by body lice), Withering felt she had hepatitis. A historian, Aronson who has published on Withering's life, has stated:

> *Withering for his part was exceptionally ungracious in his handling of the affair, and his insistence on the correctness of his behaviour in the matter seems only to have inflamed the dispute. A more judicious approach than that which he adopted might have allowed the matter to die down much more quietly.*

This is what Withering had said in reply to Darwin:

> *Possessed, therefore, as I am of self-satisfaction, of the good opinion of the world at large, and of medical men in particular, your enmity or your friendship, your good or your bad opinion, are to me equally insignificant …*

Aronson felt that Withering's handling of the affair displayed arrogance and self-satisfaction.

In October 1788, Robert Darwin was appointed physician to the Salop Infirmary. It is quite likely his father used his powers of persuasion to help bring this about.

The funding was by voluntary contributions of moneyed people, which had the advantage of allowing them to recommend patients for admission when necessary. Admissions to hospital were only on a Saturday (imagine that today!). And only between 11am and noon! One hour, with only three physicians and one surgeon. If the patient, being an honest sort said he or she could pay for the admission, it would be denied, because to allow admission would deprive the doctor of fees which he would expect if he visited the patient at home! Others denied admission were children, people with venereal disease, terminally ill patients, and pregnant women. The logic is not clear. However, casualties could be admitted at any time. Curious it is that most patients were agricultural labourers or domestic servants. No doubt their employers were anxious to facilitate their return to work, and were in a position by regular donations, to ensure this.

The Royal Salop Infirmary, 1778 (Wikipedia Commons)

Where Robert may have lacked his father's scientific genius, he was clearly well endowed in the skill of economics. His books were meticulously kept, and he made many shrewd investments. This included £1,000 at 5%

interest to rebuild St Chad's Church, whose tower had collapsed. It had been noted to have several serious cracks. Thomas Telford offered advice to repair and maintain the church, but it was ignored. Instead some local plasterers were employed to patch up the cracks. One morning at 4am the church collapsed, and a completely new church had to be constructed.

Old St Chad's Church, Shrewsbury (Courtesy of darwincountry.org)

St Chad's after the collapse in 1788 (Courtesy of darwincountry.org)

The intelligentsia in Shropshire at the time included the industrialist **William Reynolds** of Coalbrookdale, **Thomas Telford** the engineer, **Samuel Butler**, headmaster of Shrewsbury School, and **Hesba Stretton**, authoress of children's books. This was the pen name of **Sarah Smith**, (1832-1911), born in Wellington, Shropshire.

St Chad's Church after the rebuild (Geograph.org.uk)

Sarah Smith 'Hesba Stretton' (Wikipedia)

Darwin and the Wedgwoods were actively opposed to the slave trade. This carried a certain risk as it was a source of income for the well-heeled, and as such constituted a valuable part of Robert Darwin's patients. He attended a meeting in July 1791 at the Shire Hall, organised by Archdeacon Plymley. Those attending appended their names in a resolution published in the newspapers.

He also supported the formation of a local branch of the **Royal Humane Society**, founded in London in 1774, to advise the technique of resuscitation of drowning victims. This was particularly appropriate in Shrewsbury whose beautiful River Severn regularly claimed victims. Mostly these people had attempted to cross using flimsy coracles. The first local meeting was in March 1791. Robert Darwin attended together with other interested people. By the donation of one guinea, he was made a director, and committee member. There were two receiving houses equipped with first aid devices. Also, advice regarding first aid was to be placed at strategic places. The authorised apparatus and its costs were as follows:

1. A fumigator: consisting of a small pair of bellows, a brass box and handle, a flexible tube and ivory pipe	10/6
2. A spare flexible tube and pipe	2/0
3. Four wooden pipes, for blowing into the nostrils	1/6
4. Two vials of flint glass, with ground stoppers, containing spirits for smelling	1/6
5. Two flint vials, with ground stoppers, containing spirits, to be applied to the wrists etc	4/0
All these included in a box, value	4/0
Separately, a pair of bellows, for blowing into the lungs	6/0
TOTAL	£1/9/6

Quite clearly, the concept of mouth-to-mouth respiration was not entertained, and smelling salts were part of the fashion of the day.

In 1796, having become well established as a doctor, i.e. good earnings, he turned his attention to marriage. This required first and foremost, good financial planning, romance being of secondary importance. However, as a child, he had met several of the Wedgwood family, who were connected through his father's industrial interests, and visits to Stoke, (Etruria), whence he had met the Wedgwood children. Josiah Wedgwood's eldest daughter **Susannah** was his choice. Erasmus approved. During her childhood, her father's business prospered greatly.

Robert and Susannah Wedgwood were married in London in April 1796.

Shrewsbury was a centre of commerce for wool and manufactured products as well as for importing raw materials. It was connected by coach to London, a journey of 16 hours. After a couple of moves, the couple moved into their purpose built imposing residence, **The Mount**.

One might well question the cost and how it was to be met. No worries! His annual earnings were of the order of £2,000 to £3,000, his father had left him a farm in Lincolnshire, but best of all, Susannah was left a marriage settlement and will from her father of £25,000! Money enough to employ seven servants.

With regard to religion, Robert was a sceptic like his father. He copied his father's bookplate with the motto – "*E Conchis Omnia*" (Everything from snails, i.e. has evolved from more primitive forms).

However, he had to tread carefully in order not to offend his patients, many if not most, being devout believers. In regard to bringing up their children, there was a conflict between the usual disciplinarian approach, versus that propounded by **Rousseau**, whose ideas left the child without clear boundaries. As a father to his children, when he looked back to his own experiences, he must have felt confused, what with his mother (Polly) dying, his father having two children out of marriage, then re-marrying and having a further brood with Elizabeth Pole. He could be depressed, irritable and authoritarian.

Sadly, Susannah after much suffering, probably from peritonitis, died in July 1817. One hundred years later, she would no doubt have undergone abdominal surgery to elucidate the cause of her abdominal pain and vomiting. This would in all likelihood have saved her life, having revealed for instance, a possible perforated appendix, perforated stomach ulcer, or various other causes. Robert must have been particularly frustrated. He never re-married. His daughter Caroline, now aged 16, became a mother substitute for Charles, aged eight, and other sister Catherine, seven.

Meanwhile Robert, now a widower, threw himself more ardently into his work as a doctor, and business man. By way of comfort following his wife's demise, he sought solace in the pleasures of the table. According to Pattison's book, his favourite dish was goose pie:

> *Bone a large goose, then a chicken. Stuff (the chicken) inside the goose, and stuff a pickled tongue inside the chicken and place in a high standing pie crust sprinkled with beaten mace pepper and half a pound of butter.*

A Brougham (Reproduced with permission of Gail Thornton)

His height was six feet two, and weight 24 stones! When travelling, he required a specially constructed carriage, known as a Brougham, equipped with a low step. A Brougham (pronounced "broom" or "brohm") was a light, four-wheeled horse-drawn carriage built in the 19th century. It was named after Scottish jurist Lord Brougham, who had this type of carriage built to his specification by London coachbuilder Robinson & Cook in 1838 or 1839. The doctor's size and weight made home visits (which were essential in those days), somewhat precarious, as the steepness and flimsy constructions of some were clearly a potential hazard. So much so, that he would in some cases require his coachman to test the stairs before he ventured upwards. Interestingly, his only alcohol consumption was in sauces

Plaque outside The Mount

Field adjacent to The Mount

Doctor Darwin's door into the grounds of Millington's Hospital

and gravies. In regard to exercise, being a trustee of the local Millington's Hospital, he would walk down the hill, in Frankwell, (which is an area of Shrewsbury), and then ascend the other side, about one mile.

Millington's Hospital was the result of an endowment in the 1738 will of a local draper, James Millington to provide homes for 12 aged or infirm residents of Frankwell served by a common kitchen, and to provide a charity school for 20 boys and 20 girls. However, plans were not drawn up until 1747 when the land overlooking the town next to Frankwell was bought by the trustees of the will. Two barns and a beast house had to be cleared away to make room for the new building. The Palladian style chosen for the new almshouses was a fairly new style.

Despite this element of a passable lifestyle, his obesity and susceptibility to infection e.g. a carbuncle in his neck, probably due to streptococcal or staphylococcal infection, led to him resigning just before reaching the age of 65 years. Doubtless he had enough put by. It is quite possible that he had type 2 diabetes, precipitated by obesity, and which would predispose a patient to infection.

It is of interest to note that his heavy workload included a large amount of advice by letter. This today, in the author's experience, is what is called a virtual appointment. This has the advantage nowadays of distancing the patient from the doctor by dint of computer, in those days, pen and ink. The disadvantage is that only by direct personal contact can things be detected that the patient themselves may not have noticed, or attributed much importance to. Travel then was most exhausting, no comfortable car with air conditioning, contoured seats, and no ability to even out the rough road surfaces encountered. Yet despite this, he covered many miles each year. He kept an aide-memoire, and quoting from Andrew Pattison's book, for the years 1818-1824, Darwin recorded visits to Ellesmere, Wem, Myddle, Hordley in the north of the county, Newport Tern Hall (sic Hill), Broseley in the east, Kemberton, Millichope, Homer and Ludlow in the south, and Brockton in the west. He also travelled to Wales, including Welshpool, Talgarth, and Llanidloes. Plus Stafford, Lichfield, Birmingham and Herefordshire. This was a necessity because the people with money didn't live in clusters, they were scattered around the region. Having a sound business-like disposition, he carefully maintained lists of his (paying) patients. Certain names come to mind, including the Earl of Stanford (Stamford?), the Powis', the Hills, the Earl of Bradford at Weston Park, the Botfields (related by marriage to the Witherings). Considering he travelled by horse-drawn coach, in all weathers, over rutted roads, without reliable methods of heating, it must have been incredibly exhausting. It was obviously even worse than today's pot-holed roads, traffic jams, and mile upon mile of road-works. It is claimed he carried out about 1,000 home visits a year. However, he was clearly of a philanthropic nature, and consulted non-paying patients at the Salop Infirmary, as well as at his home, The Mount. Pride however then, as now, prevented some patients from seeking charity, so he devised a way around this, by asking patients to pay for the bottle, but the medicine came free. Nowadays, with the NHS, patients are given medical care, 'free at the point of delivery' only needing to pay for their drugs (unless exempt). So it might be the case that nowadays the bottle, (probably plastic) or container comes free, but the patient has to pay for its contents!

Details of patients admitted to the Salop Infirmary between midsummer 1845 and 1846, have been collated by a Mr Samuel Wood, House-Surgeon. Extracting from the data, we see that most admissions were agricultural labourers and female servants.

Disease Bronchitis Dyspepsia UlcersDisease	Job	Numbers
Bronchitis	Agricultural labourers	
Dyspepsia	"	
Ulcers	"	393 Total
Syphilis	"	
Ulcers and dyspepsia	Domestics	103
Dyspepsia	Female servants	
Abscesses and local inflammation	"	224 Total
Age range	Numbers	
10-30 years	588	
30-50 years	291	
Total (all diseases)	1059	

Robert Darwin didn't confine himself to adults, and was also sympathetic, keen and kindly with children. The patients had the whole gamut of complaints including TB, indigestion, rickets, syphilis and typhoid which he was able to recognise as different from typhus. Typhoid was spread by contaminated food or water, the latter by fleas, mites, lice or ticks. However, like today, not all patients were impressed, because his views could be somewhat outspoken, causing offence. Breaking bad news was a difficult problem then as nowadays.

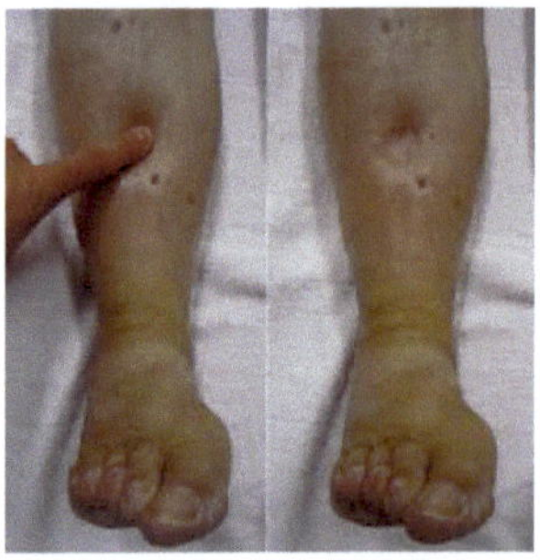

The dropsy

He also thought that an unhappy marriage could cause symptoms which had no physical cause. When it came to the dropsy, usually due to heart failure, manifested by breathlessness and oedema of the legs, he prescribed digitalis, as advised by William Withering. But unlike his father, he used smaller doses as advised by Withering, thus leading to fewer side effects. Dr Lettsom a leading physician of the time in London, had informed Withering of the side effects he had observed. Withering's reply, in calling to attention the reckless doses used by him, was truly withering:

> *Dr Lettsom has related his unsuccessful attempts' (using digitalis) … with a degree of courage and candour which do the highest honour to his integrity, but no-one can compare his choice of patients with my declarations of the fit and the unfit, or the doses he prescribed, … with my doses, rules and cautions, without being astonished that he should suppose he had been given this medicine in the manner prescribed by me, I am fully satisfied, that, had I prescribed it in such cases, such forms, such doses and such repetitions, as he has done, the effects would in my hands have been equally useless, and equally deleterious. I must, therefore, suppose that he had forgotten what I had written being conscious that his memory had deceived him. Had it been otherwise, after perusing the cases I had published … he would hardly have thought it necessary to have published more instances of what I had stigmatised as bad practice.*

Withering couldn't be accused of mincing his words!

In regard to religion, Robert was definitely a sceptic, whereas Susannah was a Unitarian. However, in order not to offend his patients, these views had to be kept in check. He shared his father's view on evolution, and

Robert Waring Darwin's bookplate

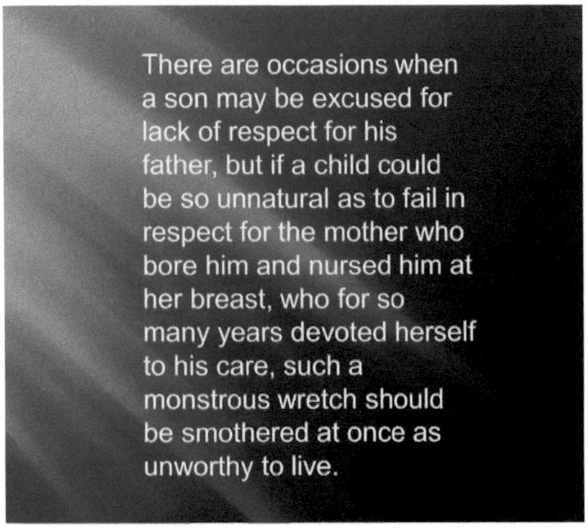

Rousseau's opinion

adopted the same motto on his bookplates, though not on his carriage, *E conchis omnia* (everything has evolved from smaller things, e.g. snails, though not literally).

These religious views came into conflict in respect of the baptism of their children. Susannah insisted her daughters be brought up in her faith, but had no objection to them also being christened in an Anglican church for the sake of their father's reputation. The family grew rapidly, four children being born within six years, which had a negative effect on their mother's health. Susannah had rheumatic fever which weakened her heart. She looked old for her age, and there were difficulties with her daughters who were described as rude and disagreeable by an aunt. This was at a time when child rearing was subjected to the progressive policy of Rousseau, one of whose aphorisms is here quoted. Yet he believed children should be allowed to express themselves as much as possible and discipline should be minimal or non-existent. Robert Darwin suffered from depression on and off, but nevertheless, according to his son Charles he was:

> *generally in high spirits and laughed and joked with everyone – often with his servants – with the utmost freedom, yet he had the art of making everyone obey him to the letter.*

(Pattison, see bibliography). Following his wife's death, he would never re-marry, a reflection of the different attitude he had compared to that of his father Erasmus, whose marital (and extra-marital) record was somewhat chequered! Naturally one would ask about the life of their children. As previously mentioned Dr Darwin was always busy, and subsequently Caroline at age sixteen became a surrogate mother to younger brother Charles.

Robert Darwin in addition to his medical skills, was a keen and astute investor, and lender of money. His areas of investment included turnpike roads and canals, including the Pontcysyllte Aqueduct carrying the Llangollen Canal.

Pontcysyllte Aqueduct (Wikipedia)

Erasmus, Robert's elder son, had been a student at Cambridge University for three years, then left for Edinburgh to study medicine. It will be recalled that his Uncle Charles, had died there whilst a medical student. So when his brother, also named Charles, joined him at Edinburgh also to study medicine, the vibes couldn't have been good. Especially because Charles wasn't all that interested in medicine. Strangely, it would come to pass that Erasmus left after a year, and although qualified, never actually practised. Charles meanwhile found the lectures dull, hated the sight of blood, and couldn't stomach watching operations carried out without anaesthetics. Possibly as a consequence of talking to his father back in Shrewsbury, he returned to Edinburgh, but to no avail. Medical practice was not for him. The alternative profession for the well-heeled was the Church, in which most of the work was left to the curate, permitting the vicar to pursue other interests, e.g. the **Reverend Edward Stone**, of

Reverend Edward Stone's blue plaque

Chipping Norton, who wrote the first scholarly article on the application of willow bark in the treatment of agues (fevers), this bark subsequently being shown to be a source of aspirin. Robert, thinking ahead, could see Charles in such a role, and thereby being able to pursue his real interest, this being natural history.

These were hard times for Robert, a widower, with sons away, and with his health failing. Charles meanwhile, at considerable expense owed to Professor John Henslow, was pursuing his interest in botany, insects, chemistry, mineralogy and geology. Unsurprisingly in what has previously been stated, Henslow was a clergyman! However, this expense proved an inspired investment, as he later offered Charles the post as naturalist on the planned **voyage of *the Beagle***. However, money remained an obstacle. Charles would have a hard job persuading his father, who had already invested heavily in him, as well as his brother. To this end, Charles used his friendship with the Wedgwood family to advantage, persuading Josiah Wedgwood (whilst out shooting on his estate), to intercede with his father. There have, over the years, been conflicting reports of Robert's views on the proposed exploratory voyage. It could easily have been construed, as what we nowadays call 'a jolly'. However, after due consideration, it would appear that Robert gave his blessing, as well as money, and the die was cast! If Erasmus Darwin was perceived as the first stage in the evolution rocket, the second stage was now on the launch pad, the astronauts being Charles and F.W. Wallace.

Charles left our coast in December 1831, and should have returned after two years. In the event, he didn't return until October 1836. The momentous events that occurred during this time will be described in the next chapter. Robert was extremely generous to all his children, an example being the farm in Lincolnshire he purchased for Charles, as well as Down House in Kent where later Charles would spend the rest of his life. By 1845, Robert was suffering frequent chest infections. His shortness of breath meant he could no longer climb stairs, (*'exertional dyspnoea'*), nor lie flat in bed (*'orthopnoea'*), and would wake at night, severely short of breath (*'paroxysmal nocturnal dyspnoea'*). Despite now nearing the inevitable, he

St Chad's Church, Montford, Shrewsbury. Robert's grave is seen in the background

The grave of Robert Waring Darwin, and his wife Susannah, Preston Montford

remained grateful for everything done for him, whispering a breathless 'thank you' on each occasion.

He died on 13th November, 1848, and the funeral was fixed for Saturday 18th. It took place at St Chad's Church, Preston Montford.

His wealth was considerable, and fairly distributed after his death. The family quite justifiably felt a deep sense of loss, as did the city of Shrewsbury. But eventually, even this would be exceeded by the fame, and achievements of Shrewsbury's son, Charles Darwin.

Chapter 5

Charles Darwin (1809-1882)

Charles Darwin (1809-1882), portrait by Ellen Wallace Sharples (Wikipedia)

Statue of Charles Darwin outside Shrewsbury Library, formerly Shrewsbury School

Charles Robert Darwin was born on 12th February 1809, in Shrewsbury, Shropshire, at his family's home, The Mount. He was the fifth of six children.

Though baptised in November 1809 at St Chad's Church, Shrewsbury, Charles and his siblings attended the Unitarian Chapel with their mother. From the age of eight, he had a taste for natural history and collecting when he joined the day school run by its preacher in 1817. Sadly his mother died in July 1817 when he was only 9yrs old. From September 1818, he joined his older brother Erasmus and they boarded at the nearby Anglican Shrewsbury School.

Robert Waring Darwin, his father, wished Charles to become a doctor, and took him along to see his poor patients in 1825. Later that year Charles went to Edinburgh Medical School, which he attended with his elder brother Erasmus. It soon became clear however, that Erasmus wasn't really interested in a career in medicine despite qualifying. Instead he moved into the social life of London in preference to the sleepy one of Shrewsbury. And their dad must have felt a sense of exasperation with his two sons' lack of enthusiasm for a career in medicine. Second best he supposed would be for Charles to become a country parson, so sent him to Christ's College, Cambridge to study for a Bachelor of Arts degree. As Charles wrote later:

> *I believe that I was considered by all my masters and by my father as a very ordinary boy, rather below the common intelligence.*

He did not even obtain an honours degree, so the current mantra to obtain a 'good' degree i.e. a 2(1), from a 'good' university, in a 'good' subject e.g. law or medicine, as apart from sociology, or media studies was to be upended by Charles Darwin's eventual achievements and universal recognition. He was especially interested in beetles and under the encouragement of one of his teachers, developed a large collection. Not only beetles, but rocks, fossils, animals, and plants. He established strategic contacts with two Cambridge professors, geologist Adam Sedgwick and botanist **John Henslow**. The latter inspired Charles, and later offered him the post of naturalist on the *Beagle*. Now there was a problem. Would his father shell out the necessary funds? And this was made especially difficult when taking into account his father's low esteem of him. His father was also worried about the dangers of such a voyage. In case his opinion was flawed, he said to Charles:

> *If you can find any man of common sense, who advises you to go, I will give my consent.*

Charles immediately thought of his Uncle Jos (Josiah) Wedgwood, who in today's parlance would be regarded as 'loaded'. As hoped, Jos persuaded Robert to accede to his son's request, and the subsequent events became history.

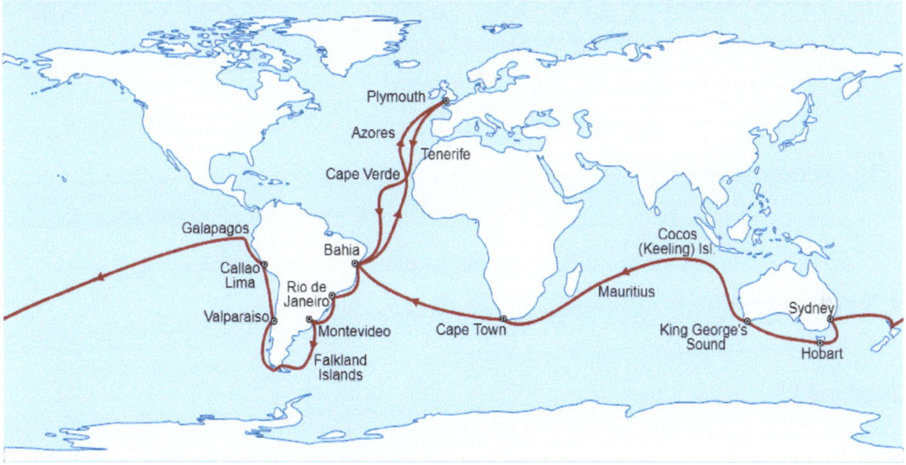

Darwin's itinerary on board the Beagle *(Wikipedia)*

So, on 27th December 1831, HMS *Beagle* left Devonport, not to return to Falmouth till 2nd October, 1836. Actually, this was the second trip made by the *Beagle*. The first was tied up with South American politics, with Britain focussing on exploration, trading and empire building. Naval stations were established in Argentina, Brazil and Chile. For safe navigation, coastal charts were required, at the behest of the British Hydrographic Office. In 1825, the Lords Commissioners of the Admiralty had ordered that two ships be prepared to survey the southern part of South America. This voyage was to end tragically with the suicide of its captain, **Pringle Stokes**, who shot himself whilst sailing in the Magellan Straits. It was ascribed to the arduous survey work and atrocious weather conditions. **Robert Fitzroy** was eventually appointed to captain the ship for its return. The purpose of a second expedition was to complete the survey of Patagonia and Tierra del Fuego, followed by the shores of Chile, Peru and other Pacific islands plus to carry out chronometric measurements around the world. After Peru, they were to set sail for the Galápagos islands. Then across the Pacific to Tahiti, New Zealand, Australia, Tasmania, Mauritius, the Cape of Good Hope, the Azores, and then back to England. Darwin recorded this as:

> *Journal of Researches into the Natural History and Geology of the Countries visited during the Voyage of H.M.S. Beagle round the World, under the command of Captain Fitz Roy, R.N.*

This is better known as the abbreviated version *The Voyage of the Beagle.*

Lest the reader imagine that the bulk of the journey was to the Galápagos, the list of his chapters, including the Galápagos, is given below. List of contents of *The Voyage of the Beagle*:

Preface	
Postscript	
I	St Jago – Cape de Verde Islands
II	Rio de Janeiro
III	Maldonado
IV	Rio Negro to Bahía Blanca
V	Bahía Blanca
VI	Bahía Blanca to Buenos Ayres
VII	Buenos Ayres to Santa Fe
VIII	Banda Oriental and Patagonia
IX	Santa Cruz, Patagonia, and the Falkland Islands
X	Tierra del Fuego
XI	Strait of Magellan – Climate of the Southern Coasts
XII	Central Chile
XIII	Chiloé and the Chonos Islands
XIV	Chiloé and Concepción – Great Earthquake
XV	Passage of the Cordillera
XVI	Northern Chile and Peru
XVII	Galápagos Archipelago
XVIII	Tahiti and New Zealand

XIX Australia
XX Keeling Island – Coral Formations
XXI Mauritius to England

Subsequently, the events in the Galápagos alone will be described. The chapter on the Galápagos was by no way the longest, in fact it amounted to just 5% of the entire text. Yet there can be no doubt, this chapter has had the greatest impact on our theory of evolution. In it, Darwin reveals an astonishing knowledge of geology, botany, zoology, and climatology. However, the first impressions on landing on Chatham Island (17th September 1835), were none too promising:

> *Nothing could be less inviting than the first appearance. A broken field of black basaltic lava, thrown into the most rugged waves, and crossed by great fissures, is everywhere covered by stunted sunburnt brushwood which shows little sign of life. One night I slept on shore on a part of the island, where black truncated cones were extraordinarily numerous.*

Black truncated cones, refers to a cone section or pyramid lacking an apex and terminating in a plane usually parallel to the base, tuff cones (sometimes called ash cones), are small monogenetic volcanic cones produced by phreatic (hydrovolcanic) explosions directly associated with magma brought to the surface through a conduit from a deep-seated magma reservoir.

Lava also flows from vents and fissures around the flanks of a volcano. As a fissure eruption progresses, the outflow tends to become localized to one or a few spots along the fissure. These spots tend to build up small cones, often referred to as parasitic cones. These parasitic cones are generally of one of two types: spatter or tuff cones.

> *The natural history of these islands is eminently curious ... most of the organic productions are aboriginal creations, found nowhere else; there is even a difference between the inhabitants of the different islands; yet*

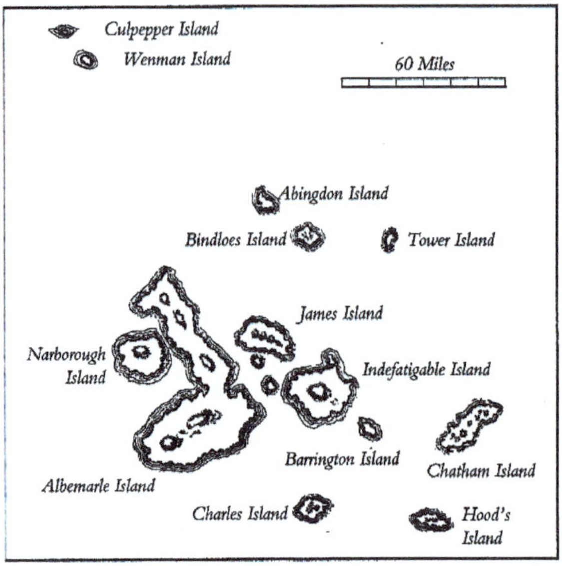

Colinpurrington.com

all show a marked relationship with those of America … though separated by (open space of ocean), between 500 and 600 miles in width.

Considering the small size of these islands, we feel the more astonished at the number of their aboriginal beings, and at their confined range. Seeing every height crowned with its crater, and the boundaries of most of the lava streams still distinct, we are led to believe that within a period, geologically recent, the unbroken ocean was here spread out. Hence, both in space and time, we seem to be brought somewhat near to that great fact – that mystery of mysteries – the first appearance of new beings on this earth.

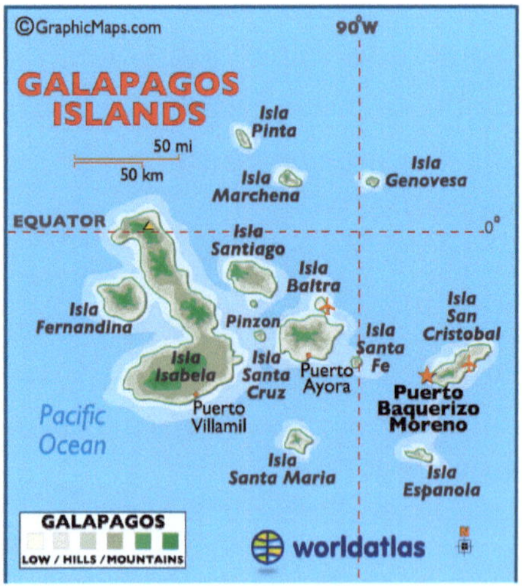

He observed aboriginality based on his discovery of species that were likely to have originated on the island in question, as applied to birds and reptiles, *compared and contrasted with their geographical separation and frequency.*

PRINCIPAL PORTS OF CALL AND STOPOVERS
(Source Dr Janet Browne, *Charles Darwin, Voyaging*, 1995)

Devonport, Plymouth	27th December, 1831
Cape Verde Islands	16th January, 1832
Bahía (San Salvador)	29th February to 18th March, 1832 and 1st to 6th August, 1836
Rio de Janeiro	5th April to 5th July, 1832
Montevideo	26th July, 1832
East Coast of South America	February, 1832 to May, 1834
Falkland Islands	March, 1833, & March, 1834
Strait of Magellan	January 1834, & May-June 1834
West coast of South America	June 1834, to September, 1835
Valparaiso	23rd July, to 10th November 1834 and 11th March to 27th April, 1835
Copiapó	22nd June to 4th July, 1835
Iquique	12th to 15th July, 1835
Callao	19th July, to 6th September, 1835
Galápagos Islands	15th September to 20th October, 1835
Tahiti	15th to 26th November, 1835
Bay of Islands, New Zealand	21st to 30th December, 1835
Sydney, Australia	12th to 30th January, 1836
Hobart, Tasmania	15th to 17th February, 1836
King George Sound, Australia	6th to 14th March, 1836
Cocos Keeling Islands	1st to 12th April, 1836
Mauritius	24th April to 9th May, 1836
Cape of Good Hope	31st May, to 15th June, 1836
St Helena	8th to 14th July, 1836
Ascension Island	19th to 23rd July, 1836
Bahía	1st to 6th August, 1836
Azores	20th to 24th September, 1836
Falmouth, Cornwall	2nd October, 1836

Spatter cones on Chatham Island (Darwin footsteps.com, with permission)

Daphne Major (D. Parer and E. Parer-Cook, with permission)

Darwin noted what he described as:

> *... a most singular group of finches, related to each other in the structure of their beaks, short tails, form of body, and plumage ... all these species are peculiar to this archipelago; and so is the whole group, with the exception of one species ... The most curious fact is the perfect gradation in the size of the beaks in the different species – from one as large as that of a hawfinch to that of a chaffinch – there are no less than six species with insensibly graduated beaks – Seeing this gradation and diversity of structure in one small intimately related group of birds in this archipelago, one species had been taken and modified for different ends.*
>
> *My attention was first thoroughly aroused, by comparing together the numerous specimens, shot by myself and several other parties on board, of the mocking-thrushes, when, to my astonishment, I discovered that all those from Charles Island belonged to one species (Mimus trifasciatus), all from Albemarle Island to M.parvulus; and all from James and Chatham Islands ... to M.melanotis ... If the different islands have*

> *their representatives of Geospiza, it may help to explain the singularly large number of the species of this sub-group in this one small archipelago ... and the perfectly graduated series in the size of their beaks.*

He went on to make the interesting observation that some of the islands were specific for being populated by a particular tortoise, this mocking thrush, these finches and numerous plants. As explanation he discounted distance of separation (Charles from Chatham a mere 50 miles and 33 from Albemarle), also the soil type, or altitude. His best explanation was the direction of ocean currents creating separations, rather than previous union because the volcanic rocks were not old enough to allow a one time union.

> *Reviewing the facts here given, one is astonished at the amount of creative force, if such an expression may be used, displayed on these small, barren, and rocky islands.*

The contrasting beaks of four Galápagos finches, three *Geospizinae* genus and one *Certhidea* (Warbler finch). Darwin originally misidentified the finches when he collected them on the Galápagos, assuming them to be from very different groups. He also failed to label them well, which he always regretted. Artist/ornithologist **John Gould** identified them for him and alerted him to them as a new group of 12 closely related species of ground finch (actually members of the Tanager family, *Thraupidae*). The name *'Darwin's finch'* was applied by **Percy Lowe** in 1936. The finches were famously studied by **David Lack** in the 1940s and more

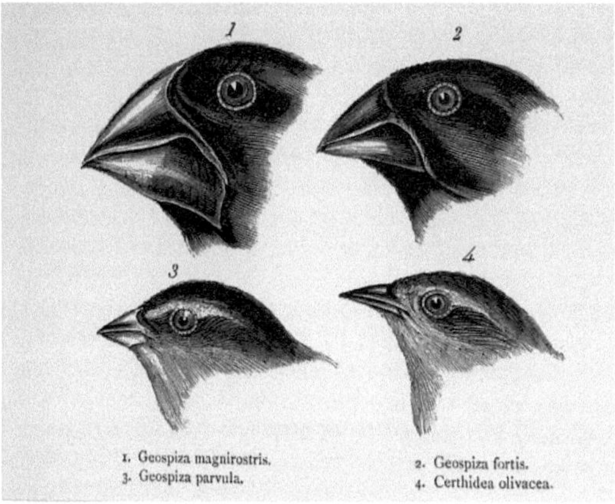

Darwin's finches (Wikipedia Commons)

recently by **Peter** and **Rosemary Grant** in 1973. The Grants provided the best evidence of evolution within a human time frame with finches found on Galápagos' Daphne Major and Minor islands.

Though easily accessible to most visitors to the Galápagos, the national park service nowadays has highly restricted visits to this island, and it is primarily used for scientific research. They examined the behaviour and life cycle of finches demonstrating the role of natural selection.

Daphne Island in 1977 was subjected to 577 days of drought. Finches that relied on small seeds and therefore were possessed of small beaks, starved to death, whereas those with larger beaks were able to survive by cracking dried out seeds courtesy of their stronger beaks. The following year, the Grants found that on average the beak sizes of the population was 3-4% larger than before the drought.

After leaving the Galápagos islands the *Beagle* continued its circumnavigation, via Tahiti, New Zealand, Tasmania, Australia, Mauritius, the Cape of Good Hope, the island of St Helena, Ascension, Bahía, the Cape Verde islands, the Azores, before finally reaching Falmouth on 2nd October, 1836. After disembarking, Darwin travelled by carriage back to Shrewsbury his family home. The next day he wrote to Captain Fitzroy:

Daphne Island (Wikipedia)

> *I arrived here yesterday morning at breakfast time, and, thank God, found my dear sisters and father quite well…Indeed all England appears changed excepting the good old town of Shrewsbury and its inhabitants*

This reaction is quite understandable taking into account all his experiences on board the *Beagle*. In his autobiography he wrote:

The Voyage of the Beagle has been by far the most important event in my life and has determined my whole career.... I have always felt that I owe to the voyage the first real training or education of my mind. I was led to attend closely to several branches of natural history; and thus my powers of observation were improved, though they were always fairly developed.

In 1837 his thoughts and ideas gleaned from the sea voyage, began to crystallize in his mind, so much so that, in his private journal he recorded:

opened first note Book on "transmutation of species". Had been greatly struck from about month of previous March on character of S.American fossils and, Species on Galápagos Archipelago. These facts (formed the sic) origin (especially later) of all my views.

Maer, where Charles Darwin married his cousin Emma Wedgwood, in January 1839

But what did he mean by *transmutation of species*?

On 29th January 1839, Darwin married his first cousin, Emma Wedgwood, Josiah Wedgwood's youngest daughter at St Peter's Church in Maer, Staffordshire.

Had Emma known of Darwin's analytical approach to marriage, she may well have had second thoughts. He composed a list on the pros and cons of marriage. The second point in the pros column stated:

Constant companion, (& friend in old age) who will feel interested in one, – object to be beloved and played with. – better than a dog anyhow.

It is unclear whether any right-minded woman today would be flattered by the last point! Eventually they would have 10 children, of whom seven

Down House (Wikipedia)

would reach adulthood. In 1842, aided by finance from his family, the Darwins moved to Down House, in the village of Downe in Kent.

Darwin at this time was not in the best of health, though we do not know from what he suffered. He enjoyed a daily exercise regimen, a quarter mile walk in the grounds of his house. He visited the spas of Moor Park, and Malvern which were fashionable at that time, though he never let the scientist in him question their claims to health promotion. In fact his daughter Annie died aged 10, and was buried in Malvern.

Darwin was elected a Member of the Royal Geological Society and the Royal Society. John Gould, an English ornithologist, reported to the Zoological Society, that he had identified a group of finches among Darwin's collection, from the Galápagos. They were very different from others, and constituted, in his opinion a special group. Within the different islands, their beaks were found to vary from strong stout, for crushing, to thin tapering, for extracting grubs, and some with a short spike to spear the food. The crucial ability to obtain food was exemplified by Malthus' views on the growth or death of populations depending on this. This was an essential part of his view on 'natural selection'. But there was a problem. A large problem. What would British Society and the Establishment (including especially the Church), make of his outlandish ideas – how could Man be descended from an ancestor shared with the apes? An infamous debate on the subject took place at Oxford in June 1860, organised by the British Association for the Advancement of Science.

Darwin did not attend, as he would doubtless have been expected to speak, a thing he disliked.

In any event, it proved a lively debate, featuring Anglican Bishop Samuel Wilberforce, (for God and creation), versus Thomas Huxley (for the idea of Man's evolution from primitive ancestors). Wilberforce was no pushover, and aimed a cannily crafted question to Huxley, to the effect that he asked whether Huxley believed himself descended from an ape on his father's or his mother's side. Huxley's response was little short of scathingly brilliant.

> 'If then, said I, the question is put to me would I rather have a miserable ape for a grandfather or a man highly endowed by nature and possessed of great means of influence and yet who employs these faculties and that influence for the mere purpose of introducing ridicule into a grave scientific discussion, I unhesitatingly affirm my preference for the ape.'

Present at the debate, and clutching his bible, was Captain Fitzroy. Darwin knew of Fitzroy's firm traditional beliefs, and was clearly uneasy about causing him any offence despite having a strong bond for his friend. Fitzroy became increasingly victim to depression, leading to his suicide by the unusual means of slitting his throat, on 30th April 1865. Darwin continued publishing on various subjects including, coral reefs, geology of South America, fossils and cirripeds (e.g. barnacles), the origin of species, effects of domestication on animals and plants, the descent of Man and selection in relation to sex, emotional expression in Man and in animals, movement in plants, and mould formation in vegetables, consequent upon the action of worms.

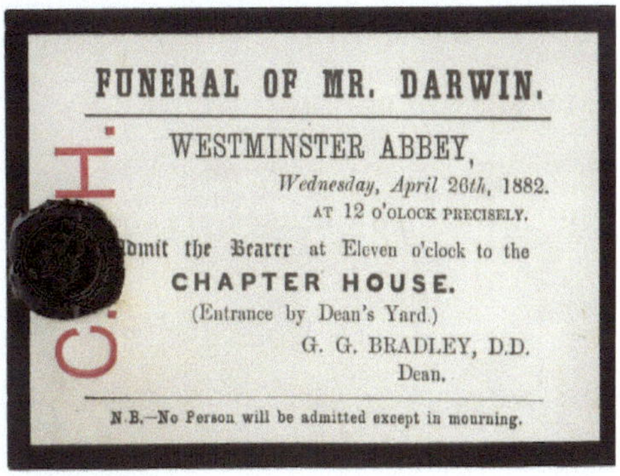
FUNERAL OF MR. DARWIN.

WESTMINSTER ABBEY,

Wednesday, April 26th, 1882.

AT 12 O'CLOCK PRECISELY.

dmit the Bearer at Eleven o'clock to the

CHAPTER HOUSE.

(Entrance by Dean's Yard.)

G. G. BRADLEY, D.D.

Dean.

N.B.—No Person will be admitted except in mourning.

C.H.

Historyofscience.com

Eventually, on 19th April 1882, with his family around him, he passed away at Down House aged 73. He was buried in Westminster Abbey.

Darwin's book *On The Origin of Species* was first published in 1859, 48 years after setting out on the *Beagle*, under the title:

> *On the Origin of Species by Means of Natural Selection or, The Preservation of Favoured Races in the Struggle for Life.*

It is amazing that, unarmed with any knowledge of genes, dominance or recessiveness, DNA, cell division by mitosis or meiosis, Darwin could write authoritatively on the origin of species, frequently interspersed with comments on varieties, though neither were defined. These matters will be covered in the next chapter. This in no way retracts from his sharp powers of observation, together with conjecture to the very limits of his active mind. Invariably this led to collision with conventional (i.e. religious) views, and later to opposition to the views of Lamarck by Darwinists. Yet in many instances, as will be seen, some of Darwin's views undoubtedly smacked of Lamarckism.

The book starts off with a description of *Variation under Domestication.* Adequacy of nutrition is regarded as a factor in maintaining 'varieties' less likely to have emerged under natural conditions. He is correct in his deduction that the way to variation in offspring resides in the reproductive element prior to conception. He attributes the heightened perception of those (breeders) who have by experience selected examples of slight variations, to the emergence of desired varieties. This will later be compared to the variability found in nature. The struggle for existence by the animal and vegetable kingdoms will be viewed from the view of Malthus, thus the production of far more than can survive under the prevailing conditions leads to the survival only of the fittest i.e. **natural selection**, and the extinction of less adapted forms, to be described in greater detail in later chapters. He admits to great difficulty in explaining transition i.e. *how a simple being or a simple organ can be changed and perfected into a highly developed being or elaboratively constructed* *organ*. Next, attention will be given to instinct by which he means the mental powers of animals. He tackles the complex subject of hybridisation which may be associated with infertility of the offspring, whereas this is not the case when varieties are crossed. The tantalizing observations resulting from incomplete geological records are covered and integrated with their geographical distribution. Finally, attempt is made with classification based on a theory of descent with modification. Also, embryology with variations appearing at different ages, and the existence of rudimentary organs.

Returning to the subject of variation under domesticity, he comments upon the contrast between the range of variation under such, and the reduced variation encountered under natural conditions. Possibly plentiful nutrition could be a factor. He quotes the French Naturalist, *Geoffroy St Hilaire* (1772-1844) when considering at what period of life variability of a species is affected, who felt that unnatural treatment of the embryo was the mechanism by which foetal anomalies could result. Geoffroy endorsed a theory of sudden evolutional change as a consequence of which these 'monstrosities' could become the founding fathers (or mothers) of new species by instantaneous transition from one form to the next.

In France between 1780 and 1815, doctors opened a broad correspondence with medical faculties and public officials about foetal anomalies which they termed *monstrosities*. Institutional and legal reforms forced doctors to encounter monstrous births with greater frequency, and they responded by developing new ideas about heredity and embryology to explain malformations to public officials. Though doctors achieved consensus on pathogenesis, they struggled to apply these ideas in forensic cases, especially with doubtful sex. Medical networks simultaneously allowed doctors to explore obstetrical techniques, as licensing regulations forced practitioners into emotional encounters with child anomalies. Doctors thus developed a new ethic for treating 'monstrosities', viewing them as pathological specimens, forensic objects, and obstetrical tragedies.

Darwin felt that these variations could be attributed to changes within the male and female reproductive systems prior to conception. He suggested that confinement may be the reason, and went on to comment how difficult it could be to breed from an animal in captivity, even when mating takes place. Similar events were found in some cultivated plants which though outwardly in good health, decline to produce seeds. He observed that carnivores, but not bears, bred well in captivity. By contrast, carnivorous birds rarely produced fertile eggs. Some domesticated animals and plants, by contrast although 'weak and sickly' bred well. Another conundrum was when wild animals though tamed and outwardly healthy, produced abnormal offspring. An exception to this, was the fertility of caged rabbits and ferrets.

Ducks provide a controversial feature, in that the domestic duck tends to have a stronger leg and weaker wing (as judged by weight of wing and leg bones) than the wild one. This is interpreted by Darwin as being the effect of its environment, not its cause. Were it so, it would support Darwinian evolutionary theory, though Darwin himself is hinting at a Lamarckian cause. Likewise the development of udders in cows and goats.

> *There are many laws regulating variation, some few of which can be dimly seen.*

One cannot but feel sorry for Darwin, the dimness of vision alluded to being the lack of genetic knowledge of the time, Mendel notwithstanding, whose seminal paper on the segregation of genes in the sweet pea would not be presented till five years after the publication of this work. Darwin's reasoning to the rules of inheritance reveal an impressive clarity of thought:

> *No breeder doubts how strong is the tendency to inheritance: like produces like is his fundamental belief; ... When a deviation appears not infrequently, and we see it in the father and child, we cannot tell whether it may be due to the same original cause acting on both; but when amongst individuals, apparently exposed to the same conditions, any very rare deviation due to some extraordinary combination of circumstances appears in the parent – say once among several million individuals – and*

> *it reappears in the child, the mere doctrine of chances almost compels us to attribute its reappearance to inheritance.*

The laws governing inheritance were to Darwin and almost everyone else, unknown – how intrigued might he have been had he known about Gregor Mendel's work. Darwin mentions situations where abnormalities are inherited by both sexes, or one sex alone. Also, that where a characteristic appears in successive generations, it seems to come earlier in subsequent generations, (a feature nowadays referred to as *genetic anticipation*). Competent judges are cited as referring to domestic animals or plants as *mere varieties*, whereas others would refer to same as the descendants of different *aboriginal species*.

The question of descent from one or several parent species is exemplified by asking it of greyhounds, bloodhounds, terriers, spaniels or bull-dogs, and if it became clear that they were all the product of one species, then great doubt about the immutability of species would be sown. Darwin professed his disbelief that all dogs were descended from one wild species, being of the opinion that they were descended from several. He felt the same with regard to cattle, but could not decide about sheep and goats. Horses he believed, albeit with some doubt, to be descended from one wild stock, and similarly poultry, ducks and rabbits. He ridiculed the opinion that many animals exhibiting minor differences had originated from different wild prototypes. Darwin introduced, but without definition, the term *race* in respect of their being created by crossing between two extremely different *races* or *species*. Great difficulties were encountered when the mongrels produced after a first cross, were themselves mated, because different features then became apparent. He would have been very much obliged to Mendel who showed how features are expressed with sweet peas, when the first generation cross (known as the F1 is itself crossed with F1) resulted in a three to one ratio of dominant colour to recessive, see the chapter on evolution and genetics. Knowledge then of the independent segregation of genes (Mendel's first and second laws) would have made the situation much clearer.

Darwin then attempted to study another group in great detail. For this purpose he chose the domestic pigeon.

> *I have kept every breed which I could purchase or obtain, ... I have associated with several eminent fanciers, and have been permitted to join two of the London pigeon clubs ... The diversity of the breeds is something astonishing. Compare the English carrier and the short-faced tumbler ...*
>
> *Great as the differences are between the breeds of pigeons, I am fully convinced that the common opinion of naturalists is correct, namely, that all have descended from the rock-pigeon ...*

Darwin gave great credit to selective breeders whose powers of observation were absolutely critical in achieving the desired result.

> *If selection consisted merely in separating some very distinct variety, and breeding from it, the principle would be so obvious as hardly to be worth notice; but its importance consists in the great effect produced by the accumulation in one direction, during successive generations, of differences absolutely inappreciable by an uneducated eye, differences which I for one have vainly attempted to appreciate.*

The same principles were applied to sheep, and horses, and many other instances. The likelihood of spotting favourable characteristics was found to be considerably greater when large numbers of the species concerned were undergoing observation.

> *Hence, in determining whether a form should be ranked as a species or a variety, the opinion of naturalists having sound judgement and wide experience seems the only guide to follow.*

He remained perplexed by the question of the status of variety, or species. The problem extended into the realm of botany. Later in the 20th century into microbiology. Here decisions were based at one time on Gram staining and biochemical reactions, e.g. whether this bacterial genus could ferment lactose e.g. *Escherichia*, or not e.g. *Salmonella*. These methods have of course been superseded nowadays by DNA and RNA analysis including ratios of guanine to cytosine ('G:C ratio').

He concluded the chapter on variation under Nature as follows (and I apologise for any confusion that follows):

> *We have, also, seen that it is the most flourishing and dominant species of the larger genera which on an average vary most; and varieties, as we shall hereafter see, tend to become converted into new and distinct species. The larger genera thus tend to become larger; and throughout nature the forms of life which are now dominant tend to become still more dominant by leaving many modified and dominant descendants. But by steps hereafter to be explained, the larger genera also tend to break into smaller genera. And thus, the forms of life throughout the universe become divided into groups subordinate to groups.*

Central to the consideration of varieties, species, and genera, is the acceptance that they follow the struggle for existence. Thus any variations that prove beneficial will generally be inherited by their offspring, and imbue them with the same advantage. This Darwin calls *natural selection*.

He carried out a simple but significant experiment. He cleared a piece of ground three feet long by two feet wide. He planted 357 seeds of weeds, but of these 295 were killed by slugs and insects. Similarly, more vigorous plants flourished at the expense of weaker ones. Another limiting factor was the preying of one species on another. But over and above this, Darwin considered climate to be the most effective of all checks. The winter of 1854-1855 killed 80% of the birds in his grounds. It can also significantly reduce numbers of various species by its effect on their food.

He went on to mention the controversial, and unfortunately topical subject of a species that was thriving due to favourable circumstances, but which could fall prey to epidemics, and here he mentioned the part of parasitic worms in animals. For animals one might substitute humans, currently in the grip of the coronavirus pandemic. In the case of crops, e.g. corn and rape-seed, though providing plenty for birds, whose numbers thereby increased considerably, yet were checked by severe winters. Comment was made on the benefits of intercrossing plant breeds compared to the ill effects of close inbreeding. As Wikipedia puts it:

> *Heterosis, hybrid vigour, or outbreeding enhancement is the improved or increased function of any biological quality in a hybrid offspring. An offspring is heterotic if its traits are enhanced as a result of mixing the genetic contributions of its parents.*

The checks and balances operating on our countryside were graphically described by his experience centred upon an estate of a relative in Staffordshire. He observed an extremely barren heath, which he compared with a similar area previously planted with Scotch fir 25 years previously. The latter presented a dramatic change in its vegetation and bird life. Near Farnham, where several heaths had been enclosed, Scotch firs had sprung up and flourished. However, where cattle had been grazing, little trees could be found, but at this juncture, the cattle had fed generously thus seeing to their fate, and no trees had survived beyond 26 years (he went to the length of checking their rings of growth!). Checks and balances also applied to the cases that he cited, e.g. insects limiting cattle in Paraguay, themselves limited by insectivorous birds, who in turn were checked by raptors or other beasts. In yet other situations, moths were central to fertilisation of orchidaceous plants and bumble bees (they were called humble bees in Darwin's time, not as a reference to their humility, but rather to the sounds they emitted!),

Red clover, Trifolium pratense *(Wikipedia)*

which fertilised the common red clover as they alone were able to reach the nectar.

However, the story doesn't end there. Because field mice destroy the combs and nests used by these bees, and the mice themselves are limited by cats.

Generally speaking, though competition between genera is great, it is surpassed by species of the same genus. (Nowadays, we have seen the grey squirrel *Sciurus carolinensis*, ousting the red *Sciurus vulgaris* in most of our island).

Unfortunately, in my view, his final sentence:

> *When we reflect on this struggle, we may console ourselves with the full belief, that the war of nature is thus incessant, that no fear is felt, that death is generally prompt, and that the vigorous, the healthy, and the happy survive and multiply –*

... is not a realistic picture of the world we currently live in. Death for instance is often preceded by prolonged suffering, at least in some instances by the incursion of medical practice, albeit with the best of intentions.

He continued with comments on sexual selection. Unlike natural selection, where the survival of the species is the predominant objective, in the case of sexual selection it is sheer lust, males competing for a female mate, without the presumable purpose (in the case of non-humans), of producing young. Should the ensuing competition result in the death of a competing male, that is unfortunate, though not necessarily the prime purpose. From a biological standpoint, the target is to improve as far as possible, the gene pool, based on the aforesaid competition. This may involve the presentation of gifts e.g. the northern cardinal bird male feeding his would-be mate, or by birdsong. Another means is by plumage, as exemplified by the peacock.

Returning to the subject of hybridisation, papilionaceous flowers (pea family), are dependent on the visit of bees for any cross fertilisation to take place. This is because the design of the space around the stamens and pistils is such that in order to obtain nectar, bees have to enter the structure intimately, and by so doing, will almost invariably brush against the stamens bearing pollen and thereby transfer it to other flowers, the pollen then fertilising the pistils.

Darwin commented that under some circumstances self-fertilisation was less successful. He compared the effect of the bees to that of a camel-hair pencil, in delicately applying the pollen. It will be apparent that in some instances, foreign pollen (i.e. from another species), could also be present in the pollen baskets found on the legs of the bee. In these cases, Darwin commented, only pollen from the same species would be successful. Another method of preventing self-fertilisation was non-contemporaneous maturation of anther and stigma.

Trees may have flowers of different sex on the same tree, as a consequence of which, although fertilisation by neighbouring flowers on the same tree is more likely, cross-fertilisation from another tree is not ruled out.

On a far larger scale, he pointed out that evolutionary change might proceed differently in a large continental mass than on islands formed as a consequence of natural events. And when in the fullness of time, the reverse takes place, the islands again becoming joined as the sea rises, the inhabitants of these erstwhile islands, which had hitherto permitted change amongst old inhabitants, would now be subjected to fierce competition on the recently modified continent. Inter-crossing of species might result in rapid change, in contrast to natural selection within a species. Natural selection results in slow changes, but to this may be added the effect of Man in selecting and nurturing those creatures manifesting features regarded as advantageous to the breeder. And concurrently, there would be some less successful species, which in competition with the more successful, would become extinct. Nevertheless, despite crowding of species, as on the Cape of Good Hope, some foreign species have become established without causing extinction of others.

Where there exists change within a species, for which the term variety applies, there is the closest competition which leads in many cases to the extinction of the less well adapted. This can apply to animals as well as to plants. Darwin in this regard, quoted the displacement in Yorkshire of the ancient black cattle, and their replacement by longhorns. On the vexed differentiation of varieties from species, he regarded them as species in the process of formation, i.e. incipient species. The cause of variation might be sheer chance. He went to some length to graphically represent the concept of variations within species leading to descendants after thousands of generations of new species, and even genera. By analogy he used the tree.

> *The affinities of all the beings of the same class have sometimes been represented by a great tree. I believe this simile speaks the truth. The green and budding twigs may represent existing species; and those produced during each former year may represent the succession of extinct species. ... Of the many twigs which flourished when the tree was a mere bush, only two or three, now grown into great branches, yet survive and bear all the other branches.*

The influence of geological changes was covered by:

> *So with the species which lived during long-past geological periods, very few now have living and modified descendants. From the first growth of the tree, many a limb and branch has decayed and dropped off; and these lost branches of various sizes may represent those whole orders, families and genera which have now no living representatives, and which are known to us only from having been found in a fossil state.*

Horse chestnut tree (branches from common ancestral branches). Photographed in the author's garden.

Darwin took it upon himself to anticipate the vexed question as to why, given there was such a diversity of species, with a presumed common ancestor, there were no in-between creatures. In part this could be answered by accepting that 'in-betweens' existed, but in time were outnumbered and superseded by more successful variants. Thus, the in-betweens although he believed they existed would be very much scarcer and unlikely to be discovered. This subject clearly vexed Darwin who accordingly turned attention to several instances where it was difficult to discern purpose. Examples he gave included the presence and advantage, if any, of webbed feet in upland geese who weren't in an aquatic environment, the route by which specialised organs such as the swim-bladders of fish appeared, the development of dischargeable electricity by some species, e.g. electric eels, the amazing complexity of the eye, the ability to detect the special senses. In all cases there was doubt nourished by the paucity of intermediate forms. He dwelled on the dictum *Natura non fecit saltum*, (Nature doesn't move by jumps). Yet the stepping stones were usually conspicuous by their absence, and there were conflicting opinions about the pace of locomotion adopted by evolution. Nevertheless, natural selection remained the abiding principle even though its fruition might only be perceived with hindsight, adaptations having taken place during 'long past periods of time'.

Darwin was clearly fascinated by the subject of instinct. Was it only obtained through inheritance, or could it be learned? He went into great detail to cite examples, often reinforced by the anecdotal communications of others whom he regarded as equally expert. He believed that domestic instincts had been acquired and natural instincts, though inherited, could by training be modified. But the Darwinian credo re. mode of inheritance goes *off piste* when he states:

> *In some cases compulsory habit alone has sufficed to produce such inherited mental changes.*

Here he mentioned dogs, cats, and chickens, whose behaviour when threatened by predators has been modified by domestication. He went to some length to mention the behaviour of the cuckoo, citing instinct as the reason for her laying her eggs in other birds' nests. Apparently the cuckoo isn't the only species of bird that does this, and he commented that American ostriches lay their eggs in several other nests, where they are hatched by males. He described in detail the slave-making instinct of certain ants, some species have workers which carry their masters, and others vice versa. The cell making instinct of hive-bees was, he found, particularly fascinating, and it still is the subject of biology lessons. It has been mathematically shown that the hexagonal shape which has evolved is the most efficient in terms of energy expenditure per gram of honey.

Returning to the subject of hybrids, Darwin went into considerable detail in regard to botany, relying in large measure on the anecdotal experiences of others e.g. Kolreuter (1773-1806) and Gartner (1772-1850), both German botanists. Generally, and after much ado with examples of plants, and animals, he ventured, albeit tentatively, to the view that if hybrids were fertile, they must have arisen from varieties, but if not, then they are probably separate species. One cannot but feel some degree of sympathy, since Darwin didn't know of Mendel's discoveries, i.e. the mere presence let alone segregation of genes, nor the existence of chromosomes, of which different species may have different numbers, though this by itself would not allow for all the observations to which he alludes. These difficulties have been compounded by the fact that the two authorities differed in their observations on a species. Darwin described hybrid vigour and was the first scientist to examine the phenomenon in a systematic manner. Hybrid vigour, or heterosis, is the increase in stature, biomass, and fertility that characterises the progeny of crosses between diverse parents such that the product is superior to the better of the two parents. In plants, this is basically achieved by a greater proliferation of cells in some but not all tissues. Inbreeding depression refers to the decline in the quantitative measure of these characters upon self-fertilisation or other forms of homozygosis of alleles (inbreeding). Although hybrids may possess vigour, this in no way is conferred on their ability to be fertile. As an example, horses and donkeys when mated produce mules of both genders. But mules when mated with other mules have very poor fertility. Yet the conclusion that hybrids invariably become sterile is not necessarily so. This Darwin quotes from the fact that aboriginal species having been successful at hybrid breeding, and that subsequent crossing whilst under domestication has been notable for good fertility. The association of sterility as a feature of separate species is clearly shaky, especially when obscured by the effects of domestication. By quoting from several sources as to the sterility or fertility of erstwhile species versus varieties, Darwin concluded that there was no fundamental distinction between species and varieties. The inherent disadvantages suffered by all these experts can nowadays be smugly explained by their observations being based almost exclusively on outward appearance, known as the *phenotype*, without knowledge of the genetic constitution, or *genotype*. The 20th century has seen the emergence of taxonomy, which in the case of bacteria relied in part on their previously described appearance after Gram staining, biochemical properties such as ability or otherwise to ferment lactose, reliance or otherwise on oxygen, later to be strengthened by genetics e.g. guanine:cytosine ('G-C') ratios, and DNA-DNA hybridisation. Thus it will be appreciated that these earlier investigators were chasing their tails when their hopes were largely based on fertility or sterility after crossing. In some ways, it is a pity Charles Darwin wasn't born 150 years later!

Having moved on away from the vexed problems of hybridism, Darwin turned his attention to clues offered by geology. Immediately the conundrum of the lack of intermediates in the apparent evolution of species was tackled with little conviction. And to increase the difficulty, he enquired into the possible age of these fossils, based on their location, and depth below the Earth's crust, then working backwards to reach a possible era. Shame, he didn't have access to carbon dating, not that that would necessarily have sufficed because carbon dating is only effective for objects less than 50,000 years old. If however the site of discovery and the nature of the Earth's crust where it was found was known, then its age could be guessed, and thus perhaps a chronological sequence of evolution might emerge. Adding to all this, specimens are often broken up into minute fragments though this is less of a problem when they were originally huge, viz. the larger dinosaurs. Unfortunately the eroding effect of ocean currents has doubtless hastened the disappearance of what could have been valuable evidence of pre-existing fossils. Oceans sink, land masses appear, oceans rise, land disappears. The eternal search for linking forms of creatures, belying the trail of evolution, is thus obscured, if indeed they ever existed. Quote from Darwin in respect of conchology:

> *This could be effected only by the future geologist discovering in a fossil state numerous intermediate gradations; and such success seems to me improbable in the highest degree.*

Belief in the transmutation of species defends its inherent incredibility by invoking the necessary enormous length of time required. Yet when we are confronted with the sudden appearance of species which runs contrary to the perceived mantra, Darwin stated that the fact would be fatal to the theory of descent with slow modification through natural selection. He admitted the weakness of his theory, but drew attention by the use of a metaphor in which the natural geological record was, as it were, an imperfect history of the world, and we were only privy to the last chapter. The perceived abrupt changes were perhaps a mirage, the explanation being that the intermediate phases had not been preserved.

Considering geographical distribution, Darwin intriguingly observed that across different continents, between similar latitudes, there was marked disparity of species, yet within any continent, regardless of latitude there was less disparity between species. This suggests independent evolution on land masses that have over time, become separated.

> *We are thus brought to the question, which has been largely discussed by naturalists namely, whether species have been created at one or more points of the earth's surface. Undoubtedly there are very many cases of extreme difficulty, in understanding how the same species could possibly have migrated from some one point to the several distant and isolated points, where now found.*

Is Darwin here guilty of loose terminology threatening to damage his theory, or did he seriously envisage creation followed by evolution by natural selection?

Migration of species was regarded as influenced by climate change, and change in the level of the land. This in turn could result in connections and disconnections of land masses, and their fauna and flora. When he explored the viability of seeds in sea water, he found that 73.5% of different kinds of seeds survived up to one month in sea-water. He also investigated fruits, green or seasoned timber which might have served as a means of transport for seeds and plants after floods. He described specific experiments:

> *Ripe hazel-nuts sank immediately, but when dried, floated for 90 days, and when planted, still germinated; similarly with asparagus; 19% of plants with ripe fruit floated after drying for 28 days.*

He concluded that 14% of plants of any country might float on the sea for more than 28 days, and still be capable of germination:

> *Going according to Johnston's Physical Atlas, the average distance covered by several Atlantic currents is 33 miles per day. On this assumption therefore the aforementioned seeds might travel across 924 miles of sea, and still be capable of germination.*
>
> *Drift timber is thrown up on most islands ... stones are embedded in the roots of trees, small parcels of earth are very frequently enclosed in their interstices ... thus completely enclosed by wood in an oak about 50 years old, three dicotyledonous plants germinated. ... the crops of floating (dead) birds contain seeds of many kinds the crops of birds do not secrete gastric juice (thus not destroying the seed).*
>
> *Earth can be found adhering to the beak and feet of birds.*

Icebergs were also cited as a means of transportation, particularly during periods of glaciation. At these times, there would be migration from the temperate to the tropical zones, then as warmth returned there would be a return to mountainous areas while excessive heat would cause extinction in the lowlands.

The relatedness of species on islands such as the Galápagos, to the nearest large land mass, i.e. South America, was said to resemble that of Cape Verde Island to Africa. Darwin's explanation was that of colonisation in the past, and definitely not independent creation. Alternatively, it could reflect earlier times when these lands were joined. But though the species appeared related, they did show minor differences, for which the explanation offered was that, though the islands of the Galápagos were close to each other, they were separated by deep ocean currents, gales were rare, thus favouring isolation. This was somewhat difficult to reconcile with the three allied species of mocking-thrushes, which, though capable of flying from island to island, appeared confined to their traditional one. This Darwin suggested, was due to pre-occupation.

Under the chapter heading, *Mutual affinities of organic beings*, Darwin stated that in the concept of descent with modification, sense appeared in the great facts in morphology. He called to attention the observation that in embryos there were parts that in adult life would be different across species, yet appeared similar at early embryonic stages. He concluded:

> *Finally, the several classes of facts which have been considered ... that the innumerable species, genera and families of organic beings, with which this world is peopled, have all descended, each within its own class or group, from common parents, and have all been modified in the course of descent, that I should without hesitation adopt this view, even if it were unsupported by other facts or arguments.*

In his *Recapitulation and Conclusion* at the end of *On The origin of Species*, Darwin has presented the final distillation of his ideas and conclusions in a carefully judged and progressive manner, such that the reader will gradually come to the acceptance of the mutability of species, even though its author at various points admits that these are his ideas, not universally accepted, and certainly not by the creationists. Their Achilles heel was that they believed it all happened within the lifetime of the Earth, which they held was of the order of 6,000 years. If evolution was Darwinian, it must be incredibly slow, and where were the intermediate forms required by the doctrine of natural selection requiring aeons of time? They reckoned without Darwin's broad mind which interrogated the world of geology and climate. With the added dimension of time, now known to be of the order of billions not thousands of years, the nascent glimmer of evolution by natural selection, gradually emerges through the mist and fog of *plans of creation*, or *unity of design*. Much verbiage is granted to those seekers after truth, in the realm of species as differentiated from varieties, and influenced by observations based on fertility. These dedicated souls are referred to as *systematists*. They will continue their pursuits whether this or that constitutes a separate species, e.g. the 50 or so British brambles, and so labelled, because of the lack of intermediate forms.

Broadening out his vision, he foresaw in the distant future, the development of psychology by distinct steps as a consequence of gradation of mental power and capacity. This would cast light on the origin of Man and his history.

He was perhaps somewhat patronising in his view of some authors of the highest eminence who seemed to be satisfied with the view that each species had been independently created. Instead he proposed all beings represented the lineal descendants of those which pre-dated the first bed of the Silurian system a geologic period that extended from the end of the Ordovician Period, about 443 million years ago, to the beginning of the Devonian Period, about 416 million years ago. He proposed that:

> *... not one living species will transmit its unaltered likeness to a distant futurity ... the species of many genera have left no descendants, but have become utterly extinct. (Only) the common and widely-spread species, belonging to the larger and dominant groups, will ultimately prevail and procreate new and dominant species.*

Fascinatingly, there have been allegations that Charles Darwin made a deathbed conversion to Christianity. According to Wikipedia:

> It was claimed by Lady Hope that Darwin said: "How I wish I had not expressed my theory of evolution as I have done." He went on to say that he would like her to gather a congregation since he "would like to speak to them of Christ Jesus and His salvation, being in a state where he was eagerly savouring the heavenly anticipation of bliss." Lady Hope's story was printed in the Boston Watchman Examiner. The story spread, and the claims were republished as late as October 1955 in the Reformation Review and in the Monthly Record of the Free Church of Scotland in February 1957.
>
> Lady Hope's story is not supported by Darwin's children. Darwin's son Francis Darwin accused her of lying, saying that "Lady Hope's account of my father's views on religion is quite untrue. I have publicly accused her of falsehood, but have not seen any reply." Darwin's daughter Henrietta Litchfield also called the story a fabrication, saying "I was present at his deathbed. Lady Hope was not present during his last illness, or any illness. I believe he never even saw her, but in any case she had no influence over him in any department of thought or belief. He never recanted any of his scientific views, either then or earlier. We think the story of his conversion was fabricated in the U.S.A. The whole story has no foundation whatsoever."

Chapter 6

Richard Owen (1804-1892)

Sir Richard Owen KCB FRMS FRS (20th July 1804 – 18th December 1892) was an English biologist, comparative anatomist and paleontologist. Despite being a controversial figure, Owen is generally considered to have been an outstanding naturalist with a remarkable gift for interpreting fossils.

Richard Owen (Wikipedia)

Owen produced a vast array of scientific work, but is probably best remembered today for coining the word *Dinosauria* (meaning "Terrible Reptile" or "Fearfully Great Reptile"). An outspoken critic of Charles Darwin's theory of evolution by natural selection, Owen agreed with Darwin that evolution occurred, but thought it was more complex than outlined in Darwin's *On the Origin of Species*. Owen's approach to evolution can be seen as having anticipated the issues that have gained greater attention with the recent emergence of evolutionary developmental biology.

Owen was born in Lancaster in 1804, one of six children of a West Indian Merchant named Richard Owen (1754-1809). He was educated at Lancaster Royal Grammar School. In 1820, he was apprenticed to a local surgeon and apothecary and, in 1824, he proceeded as a medical student to the University of Edinburgh. He left the university the following year and completed his medical course in St Bartholomew's Hospital, London, where he came under the influence of the eminent surgeon **John Abernethy**.

In July 1835 Owen married Caroline Amelia Clift in St Pancras, and by whom he had one son, William Owen. He outlived both wife and son. After his death, in 1892, he was survived by his three grandchildren and daughter-in-law Emily Owen, to whom he left much of his £33,000 fortune.

Upon completing his education, he accepted the position of assistant to William Clift, conservator of the **Museum of the Royal College of Surgeons**, on the suggestion of Abernethy. This occupation led him to abandon medical practice in favour of scientific research. He prepared a series of catalogues of the **Hunterian Collection**, in the Royal College of Surgeons and, in the course of this work, he acquired a knowledge of comparative anatomy that facilitated his researches on the remains of extinct animals.

Owen was the driving force behind the establishment, in 1881, of the **British Museum (Natural History)** in London.

In 1836, he was appointed Hunterian professor, in the Royal College of Surgeons and, in 1849, he succeeded Clift as conservator. He held the latter office until 1856, when he became superintendent of the natural history department of the British Museum. He then devoted much of his energies to a great scheme for a **National Museum of Natural History**, which eventually resulted in the removal of the natural history collections of the British Museum to a new building at South Kensington: the British Museum (Natural History) (now the **Natural History Museum**). He retained office until the completion of this work, in December, 1883, when he was made a knight of the **Order of the Bath**. He lived quietly in retirement at Sheen Lodge, Richmond Park, until his death in 1892.

Following the voyage of the *Beagle*, Darwin had at his disposal a considerable collection of specimens and, on 29th October 1836, he was

Natural History Museum, London (Wikipedia)

introduced by Charles Lyell to Owen, who agreed to work on fossil bones collected in South America.

Owen's subsequent revelations, that the extinct giant creatures were rodents and sloths, showed that they were related to current species in the same locality, rather than being relatives of similarly sized creatures in Africa, as Darwin had originally thought. This was one of the many influences that led Darwin later to formulate his own ideas on the concept of natural selection.

At this time, Owen talked of his theories, influenced by Johannes Peter Müller, that living matter had an "organising energy", a life-force that directed the growth of tissues and also determined the lifespan of the individual and of the species. Darwin was reticent about his own thoughts, understandably, when, on 19th December 1838, as secretary of the Geological Society of London, he saw Owen and his allies ridicule the Lamarckian 'heresy' of Darwin's old tutor, Robert Edmund Grant. In 1841, when the recently married Darwin was ill, Owen was one of the few scientific friends to visit; however, Owen's opposition to any hint of *transmutation* made Darwin keep quiet about his hypothesis.

Sometime during the 1840s Owen came to the conclusion that species arise as the result of some sort of evolutionary process. He believed that there was a total of six possible mechanisms: parthenogenesis, prolonged development, premature birth, congenital malformations, Lamarckian atrophy, Lamarckian hypertrophy and transmutation, of which he thought transmutation was the least likely. The historian of science Evelleen Richards has argued that Owen was likely sympathetic to developmental theories of evolution, but backed away from publicly proclaiming them after the critical reaction that had greeted the anonymously published evolutionary book *Vestiges of the Natural History of Creation* in 1844. It was revealed only decades later that the book had been authored by publisher **Robert Chambers**, (see chapter 2). Owen had been criticised for his own evolutionary remarks in his *Nature of the Limbs* in 1849. At the end of this work Owen had suggested that humans ultimately evolved from fish as the result of natural laws, which resulted in his being criticised in the *Manchester Spectator* for denying that species such as humans were created by God.

During the development of Darwin's theory, his investigation of barnacles showed, in 1849, how their segmentation related to other crustaceans, showing how they had diverged from their relatives. To both Darwin and Owen such "homologies" in comparative anatomy were evidence of descent. Owen demonstrated fossil evidence of an evolutionary sequence of horses, as supporting his idea of development from archetypes in *ordained continuous becoming* and, in 1854, gave a British Association talk on the impossibility of bestial apes, such as the recently discovered gorilla, standing erect and being transmuted into men, but Owen did not rule out the possibility that humans had evolved from other extinct animals by evolutionary mechanisms other than transmutation. To crush these ideas, Owen, as President-elect of the Royal Association, announced his authoritative anatomical studies of primate brains, claiming that the human brain had structures that apes brains did not, and that therefore humans were a separate sub-class. This started a dispute which was subsequently satirised as the *Great Hippocampus Question*. Owen's main argument was that humans have much larger brains for their body size than other mammals including the great apes. Darwin wrote that "I cannot swallow Man [being that] distinct from a Chimpanzee". The combative Thomas Henry Huxley used his March

1858 Royal Institution lecture to deny Owen's claim and affirmed that structurally, *gorillas are as close to humans as they are to baboons*. He believed that the *mental & moral faculties are essentially... the same kind in animals & ourselves*. This was a clear denial of ***Owen's claim for human uniqueness,*** given at the same venue.

On the publication of *On The Origin of Species*, Darwin sent a complimentary copy to Owen, saying *it will seem 'an abomination'*. Owen was the first to respond, courteously claiming that he had long believed that *existing influences* were responsible for the ***ordained*** birth of species. The use of this term doubtless created a spark of contention, due to its hints of religion. Darwin now had long talks with Owen who said that the book offered the best explanation *ever published of the manner of formation of species,* although he still had the gravest doubts that transmutation would bestialize Man. It appears that Darwin had assured Owen that he was looking at everything as resulting from designed laws, which Owen interpreted as showing a shared belief in *Creative Power*.

As head of the Natural History Collections at the British Museum, Owen received numerous inquiries and complaints about Darwin's book. His own views remained unknown. When emphasising to a Parliamentary committee the need for a new Natural History museum, he pointed out that:

> *The whole intellectual world this year has been excited by a book on the origin of species; and what is the consequence? Visitors come to the British Museum, and they say, 'Let us see all these varieties of pigeons: where is the tumbler, where is the pouter?' and I am obliged with shame to say, I can show you none of them ... As to showing you the varieties of those species, or of any of those phenomena that would aid one in getting at that mystery of mysteries, the origin of species, our space does not permit; but surely there ought to be a space somewhere, and, if not in the British Museum, where is it to be obtained?*

However, Huxley's attacks were making their mark. In April 1860 the *Edinburgh Review* included Owen's anonymous review of the *Origin*.

In it Owen showed his anger at what he saw as Darwin's caricature of the creationist position, and his ignoring Owen's *axiom of the continuous operation of the* ***ordained*** (this author's implementation of bold type) *becoming of living things*. As well as attacking Darwin's *disciples*, Hooker and Huxley, for their *short-sighted adherence*, he thought that the book symbolised the sort of *abuse of science... to which a neighbouring nation, some seventy years since, owed its temporary degradation* in a reference to the French Revolution. Darwin thought it *Spiteful, extremely malignant, clever, and... damaging* and later commented that:

> *The Londoners say he is mad with envy because my book is so talked about. It is painful to be hated in the intense degree with which Owen hates me.*

During the reaction to Darwin's theory, Huxley's arguments with Owen continued. Owen tried to smear Huxley, by portraying him as an "advocate of man's origins from a transmuted ape" and one of his contributions to the Athenaeum was titled "Ape-Origin of Man as Tested by the Brain". In 1862 (and on other occasions) Huxley took the opportunity to arrange demonstrations of ape brain anatomy (e.g. at the BA meeting, where William Flower performed the dissection). Visual evidence of the supposedly missing structures (*posterior cornu* and *hippocampus minor*) was used, in effect, to indict Owen for perjury. Owen had argued that the absence of those structures in apes were connected with the lesser size to which the ape brains grew, but he then conceded that a poorly developed version might be construed as present without preventing him from arguing that brain size was still the major way of distinguishing apes and humans. Huxley's campaign ran over two years and was devastatingly successful at persuading the overall scientific community, with each "slaying" being followed by a recruiting drive for the Darwinian cause. The spite lingered. While Owen had argued that humans were distinct from apes by virtue of having large brains, Huxley claimed that racial diversity blurred any such distinction. In his paper criticising Owen, Huxley directly states: "if we place A, the European brain, B, the Bosjesman (Hottentot) brain, and

C, the orang brain, in a series, the differences between A and B, so far as they have been ascertained, are of the same nature as the chief difference of those between B and C". Owen countered Huxley by saying the brains of all human races were really of similar size and intellectual ability, and that the fact that humans had brains that were twice the size of large apes like male gorillas, even though humans had much smaller bodies, made humans distinguishable. When Huxley joined the Zoological Society Council, in 1861, Owen left and, in the following year, Huxley moved to stop Owen from being elected to the Royal Society Council, accusing him "of wilful & deliberate falsehood".

In January 1863, Owen bought the *Archaeopteryx* fossil for the British Museum. It fulfilled Darwin's prediction that a proto-bird with unfused wing fingers would be found, although Owen described it unequivocally as a bird.

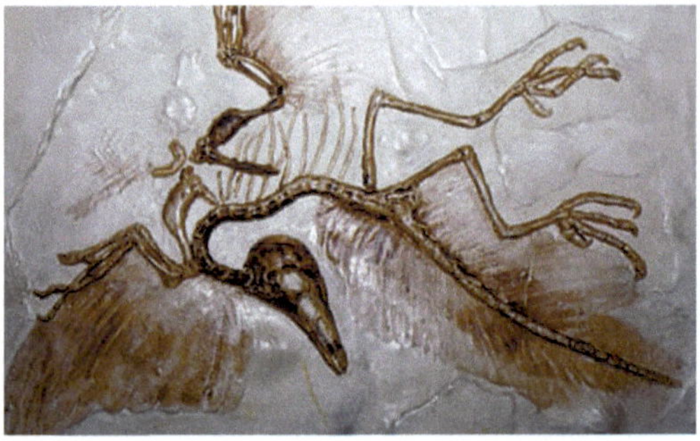

Archaeopteryx fossil (Permission from the Guardian)

The feuding between Owen and Darwin's supporters continued. In 1871, Owen was found to be involved in a threat to end government funding of Joseph Dalton Hooker's botanical collection, at Kew, possibly trying to bring it under his British Museum. Darwin commented that:

> *I used to be ashamed of hating him so much, but now I will carefully cherish my hatred & contempt to the last days of my life.*

His career was tainted by accusations that he failed to give credit to the work of others and even tried to appropriate it in his own name. This came to a head in 1846, when he was awarded the Royal Medal for a paper he had written on *belemnites* (squid like cephalopods). Owen had failed to acknowledge that the belemnite (a form of mollusc) had been discovered by Chaning Pearce, an amateur biologist, four years earlier.

As a result of the ensuing scandal, he was voted off the councils of the Zoological Society and the Royal Society.

Owen always tended to support orthodox men of science and the status quo. The royal family presented him with a cottage in Richmond Park and Robert Peel put him on the Civil List. In 1843, he was elected a foreign member of the Royal Swedish Academy of Sciences. In 1844 he became an associated member of the Royal Institute of the Netherlands. When this Institute became the Royal Netherlands Academy of Arts and Sciences in 1851, he joined as foreign member.

He died at home on 15th December 1892 and is buried in the churchyard at Ham near Richmond, Surrey.

St Andrew's Church, Ham, grave of Sir Richard Owen (1804-1892) (Wikipedia)

Owen was the first president of the Microscopical Society of London in 1839 and edited many issues of its journal – then known as *The Microscopic Journal.*

Owen also campaigned for the natural specimens in the British Museum to be given a new home. **This resulted in the establishment, in 1881, of the now world-famous Natural History Museum in South Kensington, London**. Bill Bryson argues that:

> *by making the Natural History Museum an institution for everyone, Owen transformed our expectations of what museums are for.*

By 1859, the year of the publication of Charles Darwin's *On the Origin of Species*, however, Owen's judgment was muddied by his sense that his own pre-eminence in biology was about to be lost, and he set about to discredit Darwin, who had been a good friend and colleague for 20 years. Owen wrote a very long anonymous review of the book (*The Edinburgh Review*, 1860), on which Darwin's views have already been described.

Owen is also said to have coached Bishop Wilberforce in his debate against Thomas Huxley, one of Darwin's chief defenders. As Darwin's thesis began to become more accepted in the scientific community, Owen shifted his position somewhat; although he denied Darwinian doctrine, he admitted the accuracy of its basis, claiming to have been the first to have pointed out the truth of the principle on which it was founded.

Another notorious error by Owen involved *Archaeopteryx*, the first known fossil bird, an object Owen had obtained for the museum and had described for publication in 1863. The fossil was re-examined in 1954, and scientists determined that:

> *Owen had got it upside down, dorsal for ventral, and had missed its two most important features: the breastbone, which was flat, proof that the bird could not fly but glided; and the natural cast of the brain case, which was like that of a reptile.*

AWARDS AND HONORS

Sir Richard Owen, KCB, DCL, LL.D., FRCS, FRS –

Copley Medal (1851) The Copley Medal is an award given by the Royal Society, for "outstanding achievements in research in any branch of science". It alternates between the physical and the biological sciences.

1851 Delivered the Croonian Lecture. This is a prestigious award given at the invitation of the Royal Society of London and the Royal College of Physicians. His subject, the Megatherium (Megatherium is an extinct genus of ground sloths endemic to South America that lived from the Early Pliocene through to the end of the Pleistocene).

1851 Offered a knighthood in the Order of Merit of the King of Prussia.

It is a subject of some interest how two erstwhile good friends both of immense intelligence, should have fallen out so spectacularly. And how debateable as to which left a larger mark on human knowledge. 'Evolution' and Darwin no doubt spring to mind having been conjoined for nearly two centuries (but so was Guy Fawkes and fireworks). Down House has been visited by many cognoscenti, but in no way can that compare with the legacy to mankind of London's Museum of Natural History. I leave that decision to the reader.

Chapter 7

Alfred Russel Wallace (1823-1913)

Alfred Russel Wallace (Wikipedia)

Alfred Russel Wallace was a British naturalist, explorer, geographer, anthropologist, biologist and illustrator. He is best known for independently conceiving the theory of evolution through natural selection; his paper on the subject was jointly published with some of Charles Darwin's writings in 1858. This prompted Darwin to publish *On the Origin of Species*.

Like Darwin, Wallace did extensive fieldwork; first in the Amazon River Basin, and then in the Malay Archipelago, where he identified the faunal divide now termed the *Wallace Line*, which separates the Indonesian Archipelago into two distinct parts: a western portion in which the animals are largely of Asian origin, and an eastern portion where the fauna reflects Australasia.

He was considered the 19th century's leading expert on the geographical distribution of animal species and is sometimes called the *father of biogeography*. Wallace was one of the leading evolutionary thinkers of the 19th century and made many other contributions to the development of evolutionary theory besides being co-discoverer of natural selection. These included the concepts of warning colouration in animals, and reinforcement (sometimes known as the *Wallace effect*), a hypothesis on how natural selection could contribute to speciation by encouraging the development of barriers against hybridisation. Wallace's 1904 book *Man's Place in the Universe* was the first serious attempt by a biologist to evaluate the likelihood of life on other planets. He was also one of the first scientists to write a serious exploration of the subject of whether there was life on Mars.

Wallace was strongly attracted to unconventional ideas (such as evolution). His advocacy of spiritualism and his belief in a non-material origin or the higher mental faculties of humans strained his relationship with some members of the scientific establishment.

Aside from scientific work, he was a social activist who was critical of what he considered to be an unjust social and economic system (capitalism) in 19th-century Britain. His interest in natural history resulted in his being one of the first prominent scientists to raise concerns over the environmental impact of human activity. He was also a prolific author who wrote on both scientific and social issues; and also authored an account of his adventures and observations during his explorations in Singapore, Indonesia and Malaysia. He wrote *The Malay Archipelago* first published in 1869.

Wallace had financial difficulties throughout much of his life. His Amazon and Far Eastern trips were supported by the sale of specimens he collected and, after he lost most of the considerable money he made from those sales in unsuccessful investments, he had to support himself mostly from the publications he produced. Unlike some of his contemporaries in the British scientific community such as Darwin and Charles Lyell, he had no family wealth to fall back on, and he was unsuccessful in finding a long-term salaried position and received no regular income until he was awarded a small government pension, through Darwin's efforts, in 1881.

After Wallace returned to England in 1862, he became one of the staunchest defenders of Darwin's *On the Origin of Species*. In one incident in 1863 that particularly pleased Darwin, Wallace published the short paper:

Remarks on the Rev. S. Haughton's Paper on the Bee's Cell, And on the Origin of Species

... to rebut a paper by a professor of geology at the University of Dublin that had sharply criticised Darwin's comments in the *Origin* on how hexagonal honey bee cells could have evolved through natural selection.

An even longer defence was an 1867 article in the Quarterly Journal of Science called *Creation by Law.* It reviewed the book *The Reign of Law* by George Campbell, the 8th Duke of Argyll which aimed to refute natural selection. To quote:

> *This is a common, but not a very safe argument. If the Supernatural – that is to say, the Superhuman and the Supermaterial – cannot be found nearer to us than this, it will not be securely found at all. It is very difficult to free ourselves from this notion that by going far enough back, we can "find out God" in some sense in which we cannot find Him now. The certainty not merely of one, but of many successive Creations in the history of our Planet, and especially of a time comparatively recent, when Man did not exist, is indeed an effectual answer to the notion, if it be now ever entertained, of "all things having continued as they are since the Beginning." But those who believe that the existing processes of Nature can be accounted for by "Law," may as reasonably believe that those processes were commenced by the same vague and mysterious agency. To accept the primeval narrative of the Jewish Scriptures as coming from authority, and as bringing before us the personal agency of the Creator, but without purporting to reveal the method of His work, – this is one thing. To argue that no other origin for the first parents of the human race is conceivable than that they were moulded perfect, without the instrumentality of any means, – this is quite another thing. The various hypotheses of Development, of which Darwin's theory is only a new and special version, whether they are probable or not, afford at least a possible escape from the logical puzzle which M. Guizot puts. These hypotheses are indeed destitute of proof; and in the form which they have as yet assumed, it may justly be said that they involve such*

violations of, or departures from, all that we know of the existing order of things, as to deprive them absolutely of all scientific basis. But the close and mysterious relations between the mere animal frame of Man, and that of the lower animals, does render the idea of a common relationship by descent at least conceivable.

(Certain similarities of humans and the apes renders this worthy of consideration).

Indeed, in proportion as it seems to approach nearer to processes of which we have some knowledge, it is, in a degree, more conceivable than Creation without any process, – of which we have no knowledge and can have no conception.

But whatever may have been the method or process of Creation, it is Creation still. If it were proved tomorrow that the first man was "born" from some pre-existing Form of Life, it would still be true that such a birth must have been, in every sense of the word, a new Creation. It would still be as true that God formed him "out of the dust of the earth," as it is true that He has so formed every child who is now called to answer the first question of all theologies. And we must remember that the language of Scripture nowhere draws, or seems even conscious of, the distinction which modern philosophy draws so sharply between the Natural and the Supernatural. All the operations of Nature are spoken of as operations of the Divine Mind. Creation is the outward embodiment of a Divine Idea. It is in this sense, apparently, that the narrative of Genesis speaks of every plant being formed "before it grew." But the same language is held, not less decidedly, of every ordinary birth. "Thine eyes did see my substance, yet being imperfect. In Thy book all my members were written which in continuance were fashioned, when as yet there were none of them." And these words, spoken of the individual birth, have been applied not less truly to the modern idea of the Genesis of all Organic Life. Whatever may have been the physical or material relation between its successive forms, the ideal relation has been now clearly recognised, and reduced to

> *scientific definition. All the members of that frame which has received its highest interpretation in Man, had existed, with lower offices assigned to them, in the animals which flourished before Man was born. All theories of Development have been simply attempts to suggest the manner in which, or the physical process by means of which, this ideal continuity of type and pattern has been preserved. But whilst all these suggestions have been in the highest degree uncertain, some of them violently absurd, the one thing which is certain is the fact for which they endeavour to account. And what is that fact? It is one which belongs to the world of Mind, not to the world of Matter.*

(This is a statement which draws attention to the infinite limitations of the mind as apart from the physical limitations of biology).

> *When Professor Owen tells us, for example, that certain jointed bones in the Whale's paddle are the same bones which in the Mole enable it to burrow, which in the Bat enable it to fly, and in Man constitute his hand with all its wealth of functions, he does not mean that physically and actually they are the same bones, nor that they have the same uses, nor that they ever have been, or ever can be, transferable from one kind of animal to another. He means that in a purely ideal or mental conception of the Plan of all Vertebrate skeletons, these bones occupy the same relative place – relative, that is, not to origin or use, but to the Plan or conception of that skeleton as a whole.*

(By analogy with respect to the above mentioned bones, consider the internal combustion engine as found in cars, railway engines, and piston engined aircraft. To propel on roads, or rail, or to achieve flight).

> *Here the Supermaterial, and in this sense the Supernatural, element, – that is to say, the ideal conformity and unity of conception, is the one unquestionable fact, in which we recognise directly the working of a Mind with which our own has very near relations. Here, as elsewhere,*

> *we see the Natural, in the largest sense, including and embodying the Supernatural; the Material, including the Supermaterial. No possible theory, whether true or false, in respect to the physical means employed to preserve the correspondence of parts which runs through all Creation can affect the certainty of that mental plan and purpose which alone makes such correspondence intelligible to us, and in which alone it may be said to exist.*

(Here the remarkable congruity of the properties of living creatures, is argued as evidence of their having been created. However, the process of natural selection matches this in every way, except denying purpose.)

> *It must always be remembered that the two ideas, – that of a Physical Cause and that of a Mental Purpose, – are not antagonistic only the one is larger and more comprehensive than the other. Let us take a case. In many animal frames there are what have been called "silent members" members which have no reference to the life or use of the animal, but only to the general pattern on which all vertebrate skeletons have been formed. Mr Darwin, when he sees such a member in any animal, concludes with certainty that this animal is the lineal descendant by ordinary generation of some other animal in which that member was not silent but turned to use. Professor Owen, taking a larger and wider view, would say, without pretending to explain how its presence is to be accounted for physically, that the silent member has relation to a general purpose or plan which can be traced from the dawn of Life, but which did not receive its full accomplishment until Man was born. This is certain the other is a theory. The assumed physical cause may be true or false. But in any case the mental purpose and design – the conformity to an abstract idea – this is certain. The relation in which created Forms stand to our own mind, and to our understanding of their Purpose, is the one thing which we can surely know, because it belongs to our own consciousness. It is entirely independent of any belief we may entertain, or any knowledge we may acquire, of the processes employed for the fulfilment of that Purpose.*

(There is an element of arrogance towards the concept that the ultimate achievement is Man. History is evoked in explanation, but the future is omitted, and Man is assumed to be the ultimate achievement, but this is by no means certain).

> *And yet scientific men sometimes tell us that "we must be very cautious how we ascribe intention to Nature. Things do fit into each other, no doubt, as if they were designed; but all we know about them is that these correspondences exist, and that they seem to be the result of physical laws of development and growth." Very likely; but how these correspondences have arisen, and are daily arising, is not the question, and it is immaterial how that question may be answered. Do those correspondences exist, or do they not? The perception of them by our mind is as much a fact as the sight or touch of the things in which they appear. They may have been produced by growth – they may have been the result of a process of development, – but it is none the less the development of a mental purpose. It is the end subserved that we absolutely know. What alone is doubtful and obscure is precisely that which alone we are told is the legitimate object of our research, – viz., the means by which that end has been attained. Take one instance out of millions. The poison of a deadly snake – let us for a moment consider what this is. It is a secretion of definite chemical properties which have reference, not only – not even mainly – to the organism of the animal in which it is developed, but specially to the organism of another animal which it is intended to destroy. Some naturalists have a vague sort of notion that, as regards merely mechanical weapons, or organs of attack, they may be developed by use, – that legs may become longer by fast running, teeth sharper and longer by much biting. Be it so: this law of growth, if it exist, is but itself an instrument whereby purpose is fulfilled. But how will this law of growth adjust a poison in one animal with such subtle knowledge of the organisation of another that the deadly virus shall in a few minutes curdle the blood, benumb the nerves, and rush in upon the citadel of life? There is but one explanation – a Mind, having minute and perfect knowledge of the structure of both, has designed the one to be*

> *capable of inflicting death upon the other. This mental purpose and resolve is the one thing which our intelligence perceives with direct and intuitive recognition. The method of creation, by means of which this purpose has been carried into effect, is utterly unknown.*

(Darwinian evolution would require snakes to evolve venoms of variable effectiveness at killing prey. This would require mutations, over an unknown but most likely enormous timescale. It leaves unanswered how they would survive before an effective venom evolved).

> *Perhaps no illustration more striking of this principle was ever presented than in the curious volume published by Mr Darwin on the "Fertilisation of Orchids." It appears that the fertilisation of almost all Orchids is dependent on the transport of the pollen from one flower to another by means of insects. It appears, further, that the structure of these flowers is elaborately contrived, so as to secure the certainty and effectiveness of this operation. Mr Darwin's work is devoted to tracing in detail what these contrivances are. To a large extent they are purely mechanical, and can be traced with as much clearness and certainty as the different parts of which a steam-engine is composed. – The complication and ingenuity of these contrivances almost exceed belief. "Moth – traps and spring – guns set on these grounds," might be the motto of the Orchids. There are baits to tempt the nectar loving Lepidoptera, with rich odours exhaled at night, and lustrous colours to shine by day; there are channels of approach along which they are surely guided, so as to compel them to pass by certain spots; there are adhesive plasters nicely adjusted to fit their probosces, or to catch their brows; there are hair-triggers carefully set in their necessary path, communicating with explosive shells, which project the pollen-stalks with unerring aim upon their bodies. There are, in short, an infinitude of adjustments, for an idea of which I must refer my readers to Mr Darwin's inimitable powers of observation and description – adjustments all contrived so as to secure the accurate conveyance of the pollen of the one flower to its precise destination in the structure of another.*

Now there are two questions which present themselves when we examine such a mechanism as this. The first is, "What is the use of the various parts, or their relation to each other with reference to the purpose of the whole?" The second question is, "How were those parts made, and out of what materials?" It is the first of these questions – that is to say, the use, object, intention, or purpose of the different parts of the plant, which Darwin sets himself instinctively to answer first; and it is this which he does answer with precision and success. The second question, that is to say, how those parts came to be developed, and out of what "primordial elements" they have been derived in their present shapes, and converted to their present uses – this is a question which Darwin does also attempt to solve, but the solution of which is in the highest degree difficult and uncertain. It is curious to observe the language which this most advanced disciple of pure naturalism instinctively uses when he has to describe the complicated structure of this curious order of plants. "Caution in ascribing intentions to nature" does not seem to occur to him as possible. Intention is the one thing which he does see, and which, when he does not see, he seeks for diligently until he finds it. He exhausts every form of words and of illustration by which intention or mental purpose can be described. "Contrivance" "curious contrivance" – "beautiful contrivance," – these are expressions which recur over and over again. Here is one sentence describing the parts of a particular species; "the Labellum is developed into a long nectary, in order to attract Lepidoptera, and we shall presently give reasons for suspecting that the nectar is purposely so lodged that it can be sucked only slowly, in order to give time for the curious chemical quality of the viscid matter setting hard and dry." Nor are these words used in any sense different from that in which they are applicable to the works of man's contrivance – to the instruments we use or invent for carrying into effect our own preconceived designs. On the contrary, human instruments are often selected as the aptest illustrations both of the object in view, and of the means taken to effect it. Of one particular structure Mr Darwin says: – "This contrivance of the guiding ridges may be compared to the little instrument sometimes used for guiding a thread into the eye of a

> *needle." Again, referring to the precautions taken to compel the insects to come to the proper spot, in order to have the "pollinia" attached to their bodies, Mr Darwin says: "Thus we have the rostellum partially closing the mouth of the nectary, like a trap placed in a run for game, and the trap so complex and perfect!" But this is not all. The idea of special use, as the controlling principle of construction, is so impressed on Mr Darwin's mind, that, in every detail of structure, however singular or obscure, he has absolute faith that in this lies the ultimate explanation. If an organ is largely developed, it is because some special purpose is to be fulfilled.*

This assumes that purpose 'ultimate explanation' has a part to play, as apart from sheer randomness.

> *If it is aborted or rudimentary, it is because that purpose is no longer to be subserved. In the case of another species whose structure is very singular, Mr Darwin had great difficulty in discovering how the mechanism was meant to work, so as to effect the purpose. At last he made it out, and of the clue which led to the discovery he says: – "The strange position of the Labellum perched on the summit of the column, ought to have shown me that here was the place for experiment. I ought to have scorned the notion that the Labellum was thus placed for no good purpose. I neglected this plain guide, and for a long time completely failed to understand the flower."*

It is likely that in the above example, the Labellum (in botany, the labellum is the part of the flower of an orchid or Canna, or other less-known genera, that serves to attract insects, which pollinate the flower, and acts as a landing platform for them. Labellum is the Latin diminutive of labrum, meaning lip) was a design which had a purpose in one variety of plant, but no apparent purpose in another. Darwin has suggested it was part of an experiment, and presumably the result would be its effectiveness or non-effectiveness in attracting insects for pollination. How would the result of this 'experiment' become apparent?

When we come to the second part of Mr Darwin's work, viz., the Homology of the Orchids, we find that the inquiry divides itself into two separate questions, – first, the question what all these complicated organs are in their primitive relation to each other; and secondly, how these successive modifications have arisen, so as to fit them for new and changing uses. Now, it is very remarkable that of these two questions, that which may be called the most abstract and transcendental – the most nearly related to the Supernatural and the Supermaterial – is again precisely the one which Darwin is able to solve most clearly. We have already seen how well he solves the first question – What is the use and intention of these various parts? The next question is, What are these parts in their primal order and conception? The answer is, that they are members of a numerical group, having a definite and still traceable order of symmetrical arrangement. They are expressions of a numerical idea, as so many other things – perhaps as all things – of beauty are. Mr Darwin gives a diagram, showing the primordial or archetypal arrangement of Threes within Threes, out of which all the strange and marvellous forms of the Orchids have been developed, and to which, by careful counting and dissection, they can still be ideally reduced. But when we come to the last question – By what process of natural consequence have these elementary organs of Three within Three been developed into so many various forms of beauty, and made to subserve so many curious and ingenious designs? – we find nothing but the vaguest and most unsatisfactory conjectures. Let us take one instance as an example. There is a Madagascar Orchis – the "Angro – cum sesquipedale" – with an immensely long and deep nectary. How did such an extraordinary organ come to be developed? Mr Darwin's explanation is this. The pollen of this flower can only be removed by the proboscis of some very large Moths trying to get at the nectar at the bottom of the vessel. The Moths with the longest probosces would do this most effectually; they would be rewarded for their long noses by getting the most nectar; whilst, on the other hand, the flowers with the deepest nectaries would be the best fertilised by the largest Moths preferring them. Consequently, the deepest-nectaried Orchids, and the longest – nosed Moths, would each confer on the other a great advantage in the "battle of life". This would tend to their respective perpetuation, and to the constant lengthening of nectaries and of noses. But the passage is so curious and characteristic, that it is well to give Mr Darwin's own words:

"As certain Moths of Madagascar became larger, through natural selection in relation to their general conditions of life, either in the larval or mature state, or as the proboscis alone was lengthened to obtain honey from the Angraecum, those individual plants of the Angraecum which had the longest nectaries (and the nectary varies much in length in some Orchids) and which, consequently, compelled the Moths to insert their probosces up to the very base, would be the best fertilised. These plants would yield most seed, and the seedlings would generally inherit longer nectaries; and so it would be in successive generations of the plant and Moth. Thus it would appear that there has been a race in gaining length between the nectary of the Angraecum and the proboscis of certain Moths; but the Angraecum has triumphed, for it flourishes and abounds in the forests of Madagascar, and still troubles each Moth to insert its proboscis as far as possible in order to drain the last drop of nectar ... We can thus," says Mr Darwin, "partially understand how the astonishing length of the nectary may have been acquired by successive modifications."

It is indeed but a "partial" understanding. How different from the clearness and the certainty with which Mr Darwin is able to explain to us the use and intention of the various organs! Or the primal idea of numerical order and arrangement which governs the whole structure of the flower! It is the same through all Nature. Purpose and intention, or ideas of order based on numerical relations, are what meet us at every turn, and are more or less readily recognised by our own intelligence as corresponding to conceptions familiar to our own minds. We know, too, that these purposes and ideas are not our own, but the ideas and purposes of Another of One whose manifestations are indeed superhuman and supermaterial, but are not "supernatural", in the sense of being strange to Nature, or in violation of it.

> *The truth is, that there is no such distinction between what we find in Nature, and what we are called upon to believe in Religion, as that which men pretend to draw between the Natural and the Supernatural. It is a distinction purely artificial, arbitrary, unreal. Nature presents to our intelligence, the more clearly the more we search her, the designs, ideas, and intentions of some:*
>
> *"Living Will that shall endure,*
> *When all that seems shall suffer shock."*
>
> *Religion presents to us that same Will, not only working equally through the use of means, but using means which are strictly analogous – referable to the same general principles – and which are constantly appealed to as of a sort which we ought to be able to appreciate, because we ourselves are already familiar with the like. Religion makes no call on us to reject that idea, which is the only idea some men can see in Nature – the idea of the universal Reign of Law – the necessity of conforming to it – the limitations which in one aspect it seems to place on the exercise of Will, the essential basis, in another aspect, which it supplies for all the functions of Volition. On the contrary, the high regions into which this idea is found extending, and the matters over which it is found prevailing, is one of the deepest mysteries both of Religion and of Nature. We feel sometimes as if we should like to get above this rule – into some secret Presence where its bonds are broken. But no glimpse is ever given us of anything, but "Freedom within the bounds of Law". The Will revealed to us in Religion is not – any more than the Will revealed to us in Nature – a capricious Will, but one with which, in this respect, "there is no variableness, neither shadow of turning."*

(Curious that if we mentally substitute nectaries for highly placed leaves, and long probosces for long necks, we might recall Lamarck's giraffe!)

After an 1870 meeting of the British Science Association, Wallace wrote to Darwin complaining that there were "no opponents left who know anything of natural history, so that there are none of the good discussions we used to have."

Historians of science have noted that, while Darwin considered the ideas in Wallace's paper to be essentially the same as his own, there were differences. Darwin emphasised competition between individuals of the same species to survive and reproduce, whereas Wallace emphasised environmental pressures on varieties and species forcing them to become adapted to their local conditions, leading populations in different locations to diverge. Some historians, have suggested the possibility that in the paper he mailed to Darwin, Wallace did not discuss selection of individual variations but group selection.

Others have noted that another difference was that Wallace appeared to have envisioned natural selection as a kind of feedback mechanism keeping species and varieties adapted to their environment (now called 'stabilizing', as opposed to 'directional' selection). They point to a largely overlooked passage of Wallace's famous 1858 paper:

> *The action of this principle is exactly like that of the centrifugal governor of the steam engine, which checks and corrects any irregularities almost before they become evident; and in like manner no unbalanced deficiency in the animal kingdom can ever reach any conspicuous magnitude, because it would make itself felt at the very first step, by rendering existence difficult and extinction almost sure soon to follow.*

Scholars have continued to explore the connection between natural selection and systems theory, which is the interdisciplinary study of systems. A system is a cohesive conglomeration of interrelated and interdependent parts which can be natural or man-made. Warning colouration was one of a number of contributions by Wallace in the area of the evolution and its role in protection. It was also a lifelong disagreement with Darwin about the importance of sexual selection.

In 1867, Darwin wrote to Wallace about a problem in explaining how some caterpillars could have evolved conspicuous colour schemes. Darwin had come to believe that many conspicuous animal colour schemes were due to sexual selection. However, this could not apply to caterpillars.

Wallace responded that he and Henry Bates had observed that many of the most spectacular butterflies had a peculiar odour and taste, and that he had been told by John Jenner Weir that birds would not eat a certain kind of common white moth because they found it unpalatable.

It seemed likely that the conspicuous colours served as a warning to predators and thus could have evolved through natural selection. Darwin was impressed by the idea. At a later meeting of the Entomological Society, Wallace asked for any evidence anyone might have on the topic. In 1869, Weir published data from experiments and observations involving brightly coloured caterpillars that supported Wallace's idea.

Wallace attributed less importance than Darwin to sexual selection. In his 1878 book *Tropical Nature and Other Essays*, he wrote extensively about the colouration of animals and plants and proposed alternative explanations for a number of cases Darwin had attributed to sexual selection.

In 1889, Wallace wrote the book *Darwinism*, which explained and defended natural selection. In it, he proposed the hypothesis that natural selection could drive the reproductive isolation of two varieties by encouraging the development of barriers against hybridisation. Thus it might contribute to the development of new species. He suggested the following scenario:

When two populations of a species had diverged beyond a certain point, each adapted to particular conditions, hybrid offspring would be less adapted than either parent form and so natural selection would tend to eliminate the hybrids. Furthermore, under such conditions, natural selection would favour the development of barriers to hybridisation, as individuals that avoided hybrid matings would tend to have more fit offspring, and thus contribute to the reproductive isolation of the two incipient species.

This idea came to be known as the *Wallace effect*, later termed *reinforcement*. Wallace had suggested to Darwin that natural selection could play a role in preventing hybridisation in private correspondence as early as 1868, but had not worked it out to this level of detail.

It continues to be a topic of research in evolutionary biology today, with both computer simulation and empirical results supporting its validity. (However, it is at odds with the experience of so called *hybrid vigour* or *Heterosis,* which results in the improved or increased function of any biological quality in a hybrid offspring. An offspring is heterotic if its traits are enhanced as a result of mixing the genetic contributions of its parents).

The Origin of Species,
and the Sarawack Law

1855
On the Law Which Has Regulated
The Introduction Of New Species
By Alfed Russell Wallace

February 1855
(Also Known As The Sarawack Law)

Wallace's views in this area were at odds with two major tenets of the emerging Darwinian philosophy, which were that evolution was not teleological (purpose driven) and that it was not anthropocentric (human-centred). Much later in his life Wallace returned to these themes, that evolution suggested that the universe might have a purpose and that certain aspects of living organisms might not be explainable in terms of purely materialistic processes, in a 1909 magazine article entitled *The World of Life*, which he later expanded into a book of the same name.

In many accounts of the development of evolutionary theory, Wallace is mentioned only in passing as simply being the stimulus to the publication of Darwin's own theory. In reality, Wallace developed his own distinct evolutionary views which diverged from Darwin's, and was considered by many (especially Darwin) to be a leading thinker on evolution in his day, whose ideas could not be ignored. One historian of science has pointed out that, through both private correspondence and published works, Darwin and Wallace exchanged knowledge and stimulated each other's ideas and theories over an extended period. Wallace is the most-cited naturalist in Darwin's *Descent of Man*, occasionally in strong disagreement.

Wallace remained an ardent defender of natural selection for the rest of his life. By the 1880s, evolution was widely accepted in scientific circles, but natural selection less so. Wallace's book, *Darwinism*, was published as a response to the scientific critics of natural selection. Of all Wallace's books, it is the most cited by scholarly publications.

He also wrote poetic verse, an example being *A Description of Javita* from his book *Travels on the Amazon*.

The poem begins:

> 'Tis where the streams divide, to swell the floods
> Of the two mighty rivers of our globe;
> Where gushing brooklets in their narrow beds
> There is an Indian village; all around,
> The dark, eternal, boundless forest spreads
> Its varied foliage. Stately palm-trees rise
> On every side, and numerous trees unknown
> Save by strange names uncouth to English ears.
> Here I dwelt awhile the one white man
> Among perhaps two hundred living souls.
> They pass a peaceful and contented life
> I'd be an Indian here, and live content
> To fish, and hunt, and paddle my canoe,
> And see my children grow, like young wild fawns,
> In health of body and in peace of mind,
> Rich without wealth, and happy without gold!

On 7th November 1913, Wallace died at home in the country house he called *Old Orchard*, which he had built a decade earlier. He was 90 years old and his death was widely reported in the press. The *New York Times* called him:

Wallace's grave in Broadstone Cemetery, Dorset, UK which was restored by the A. R. Wallace Memorial Fund in 2000. It features a 7-foot (2.1 m) tall fossil tree trunk from Portland mounted on a block of Purbeck limestone (Wallace Memorial Fund)

the last of the giants belonging to that wonderful group of intellectuals that included, among others, Darwin, Huxley, Spencer, Lyell, and Owen, whose daring investigations revolutionised and evolutionised the thought of the century.

Some of Wallace's friends suggested that he be buried in Westminster Abbey, but his wife followed his wishes and had him buried in the small cemetery at Broadstone, Dorset. Several prominent British scientists formed a committee to have a medallion of Wallace placed in Westminster Abbey near where Darwin had been buried. The medallion was unveiled on 1st November 1915.

The impact of Wallace's views on evolution owed much to the fact that he shared them with Darwin. This was no doubt due to English society of the time. Thus the joint reading of their papers on natural selection associated Wallace with the more famous Darwin, although it must be said some of Wallace's iconoclastic opinions in the ensuing years would test Darwin's patience. Nevertheless, they remained on friendly terms for the rest of Darwin's life, despite Wallace's interest in spiritualism.

When evolution is conceived in most people's minds, it is perceived as involving a progressive increase in cell number and complexity, whereas this need not necessarily be the case. Most organisms have adapted to specific environments. Multicellular life is probably not the goal of evolution. The developmental sense of evolution, has traditionally been referred to as the *scala naturae*, popular with science fiction screenwriters. The next species that evolves from humans may be a herbivorous and apelike animal that has adapted to global warming.

The Darwin-Wallace medal, 1908, issued by the Linnaean Society on the 50th anniversary of their joint paper on natural selection

Darwin had the view that choice of mate in the animal world could be affected by its *colour*, but this proved a problem of concept in regard to caterpillars, since they were not at a stage in which choice of mate would be relevant. Wallace responded that many of the most spectacular butterflies had a peculiar *odour* and taste, and that birds would not eat a certain kind of common white moth because they found it unpalatable.

> *Now, as the white moth is as conspicuous at dusk as a coloured caterpillar in the daylight.*

Wallace wrote back to Darwin that it seemed likely that the conspicuous colour scheme served as a warning to predators and thus could have evolved through natural selection. Darwin was impressed by the idea. **Warning colouration** was one of a number of contributions Wallace made in the area of the evolution of animal colours in general and the concept of protective colouration in particular.

Wallace Effect
1889: Darwinism
When two populations of a species have diverged, hybrid offspring would be less-well adapted and natural selection would eliminate the hybrids.
Individuals who avoid hybridisation (for whatever reason) would be fitter.

The liberal clergy accepted the concept of natural selection as God's design, **Baden Powell** (later Lord Baden Powell) who founded the Boy Scouts' Movement, stated that explanation of events as miracles was in fact atheistic, because God's law was the use of evolution, not miracles. Science was strengthening its hold, despite not removing God.

Darwin believed humans were descended from 'earlier animals' and his theory extended to the evolution of society and intelligence. He included the observation of decorative beauty in wildlife. Thus, in order to ensure cross fertilisation in orchids, their flowers were adapted in order to attract specific moths. This supported his ideas about natural selection. But in regard to the mechanism of heredity, as applied to observations on the variation of domestic animals and plants, he suggested *pangenesis i.e.* that each part of the body continually emitted its own type of small organic particles called *gemmules* that aggregated in the testes or ovaries contributing heritable information to the sperm or eggs.

The idea of *transmutation of species* was propounded by Erasmus Darwin, Lamarck, Saint-Hilaire and Grant. These views were attacked by others including Lyell who believed that species were immutable. Geographical separation was believed by Wallace to have a role in the divergence of species. He asked the question:

> *Are very closely allied species ever separated by a wide interval of country?*

Observations regarding the geographic and geologic distribution of both living and fossil species, would become known as biogeography. His conclusion that:

> *Every species has come into existence coincident both in space and time with a closely allied species*

has come to be known as the *Sarawak Law.* In 1864, Wallace published a paper titled, *The Origin of Human Races and the Antiquity of Man Deduced from the Theory of Natural Selection.* He explained the apparent stability of the human stock by pointing to the vast gap in cranial capacities between humans and the great apes. He saw the evolution of humans in two stages: achieving a bipedal posture freeing the hands to carry out the dictates of the brain, and the *recognition of the human brain as a totally new factor in the history of life.* Wallace was apparently the first evolutionist to recognise clearly that ... *with the emergence of that bodily specialization which constitutes the human brain, bodily specialization itself might be said to be outmoded.* For this paper he won Darwin's praise. Strangely afterwards, Wallace became a spiritualist. He was unable to understand how natural selection could account for mathematical, artistic, or musical genius, as well as metaphysical musings, wit and humour. Surely something in *the unseen universe of Spirit* had interceded at least three times in history. The first was the creation of life from inorganic matter. The second was the introduction of consciousness in the higher animals. And the third was the generation of the higher mental faculties in humankind. His spiritualist belief led him to believe that the purpose of the universe was the development of the human spirit. These views greatly disturbed Darwin, who felt that spiritual appeals were not necessary and that sexual selection could easily explain apparently non-adaptive mental phenomena. While some historians have concluded that Wallace's belief that natural selection was insufficient to explain the development of consciousness and the human mind was directly caused by his adoption of spiritualism, other scholars have disagreed, and some maintain that Wallace never believed natural selection applied to those areas. Reaction to Wallace's ideas on this topic among leading naturalists at the time varied. Charles Lyell endorsed Wallace's views on human evolution rather than Darwin's. Wallace's belief that human consciousness could not be entirely a product of purely material causes was shared by a number of prominent intellectuals in the late 19th and early 20th centuries. However, many, including Huxley, Hooker, and Darwin himself, were critical of Wallace. Darwin did not believe that evolution was purpose driven, or centred on Man. Much later in his life Wallace returned to these themes, that evolution suggested the universe might have a purpose and that certain aspects of living organisms might not be explainable in terms of purely materialistic processes. If materialistic notions fail, the vacuum is rapidly filled by some form of creation unencumbered by religion, or by some religion of historical interest, or another not yet experienced. Wallace's views diverged from those of Darwin who nevertheless regarded him as a leading thinker of the time.

Darwin and Wallace both agreed on the existence of natural selection but whereas Wallace believed that evolution had a purpose *(teleology)* in maintaining species' fitness to their environment, Darwin failed to be drawn of the idea of purpose, instead feeling it just happened so.

Wallace crossed the narrow strait between the islands of Bali and Lombok in 1856. On the north coast of Bali he saw birds highly characteristic of Asian ornithology and expected to see the same birds as soon as he crossed the Lombok Strait. Amazingly on the shores of Lombok he never saw the same birds again. He found a totally different set of species, namely white cockatoos and honeyeaters which are characteristic of Australia and are entirely absent from the western region of the archipelago. He wrote in his book, *The Australian Archipelago:*

> *The great contrast between the two divisions of the Archipelago is nowhere so abruptly exhibited as on passing from the island of Bali to that of*

> *Lombok, where the two regions are in closest proximity ... The strait is here fifteen miles wide, so that we may pass in two hours from one great division of the earth to another, differing as essentially in their animal life as Europe does from America.*

The Lombok Strait represents part of the **biogeographical** boundary between the fauna of Asia and those of Australasia which was subsequently named the **Wallace Line**. On the Asian side of the Wallace Line are the Asian elephant, the rare Javanese rhinoceros, Sumatran tigers and Borneo leopards, all kinds of monkeys, the orangutans of Sumatra and Borneo, and numerous birds that are specific to Asia. On the Australasian side are the marsupials such as the possum-like cuscus and the tree kangaroos, as well as birds specific to Australasia such as white cockatoos, honeyeaters, brush turkeys and the spectacular Birds of Paradise. By his observations Alfred Russel Wallace had made a major contribution to a new science, the science of **biogeography**, or of the relationship between zoology and geography.

Wallace did not confine himself to the biogeography of living species, but also included evidence from the fossil record to discuss the processes of evolution and migration that had led to the geographical distribution of modern animal species. He initially believed the most likely cause for the rapid extinctions was glaciation, but later felt they were *due to man's agency*.

Sumatran tiger. Found west of the Wallace Line (Wikipedia)

Australian cockatoo. Found east of the Wallace Line

BALI Wallace Line

The Wallace Line

The appearance of species, *Speciation*, can occur when a part of the population becomes *geographically* isolated from the main population. In time minor changes ensue some of which confer survival advantage, e.g. beak size and shape. The population becomes entirely separated and finally constitutes a new species. Thus geographic isolation brings about *allopatric speciation*. An important example of this type of speciation is formation of Darwin's finches that formed separate species in the Galápagos Islands. A different mechanism occurs when the reproductive system itself undergoes change such that creatures altered in this manner, can no longer breed with the original ones, although not geographically separate. This is known as *sympatric* speciation, an example of which are two types of Orcas found in the Pacific Ocean. *Parapatric* speciation is the case when some members of the original species become isolated, and enter a new niche or habitat. Some grasses have adapted to the soil contaminated by heavy metals in the vicinity of mines. Yet another type of speciation is *Quantum* and refers to the *rapid* and *abrupt* mode of species formation defined as:

> *the budding of a new and very different daughter species from a semi-isolated peripheral population of the ancestral species.*

This type of speciation is based on the observation on *Drosophila* (fruit flies) inhabiting Hawaii Island.

Wallace classified islands into three different types:

Oceanic, e.g. the Galápagos, and Hawaiian, because they were formed in mid-ocean, and had never been part of a continent. They consequently lacked any terrestrial animals, or amphibians, having been accidentally colonised by them and these had subsequently evolved.

Continental (recently separated) e.g. Britain.

Continental (long time separated) e.g. Madagascar, e.g. lemurs, which are remnants of once widespread continental faunas. He also included the possible effects of previous Ice Ages.

Wallace was aware of the impact of human activities on the natural world and he warned about the dangers of deforestation and soil erosion, especially in tropical climates prone to heavy rainfall. Noting the complex interactions between vegetation and climate, he warned that the extensive clearing of rainforest for coffee cultivation in Ceylon, nowadays Sri Lanka, and India would adversely impact the climate in those countries and lead to their eventual impoverishment due to soil erosion. These comments are especially poignant today, when we hear of deforestation and burning to make way for palm oil cultivation. Quoting Wallace:

> *These considerations should lead us to look upon all the works of nature, animate or inanimate, as invested with a certain sanctity, to be used by us but not abused, and never to be recklessly destroyed or defaced. To pollute a spring or a river, to exterminate a bird or beast, should be treated as moral offences and as social crimes; ... Yet during the past century, which has seen those great advances in the knowledge of Nature of which we are so proud, there has been no corresponding development of a love or reverence for her works; so that never before has there been such widespread ravage of the earth's surface by destruction of native vegetation and with*

> *it of much animal life, and such wholesale defacement of the earth by mineral workings and by pouring into our streams and rivers the refuse of manufactories and of cities.*

But long before the Darwins and Wallaces, Buffon had challenged conventional i.e. religious views on evolution. He was of the belief that the Earth was far older than the 6,000 years held by religious doctrine. Instead of seven days in the making of earth, he proposed seven epochs. Faced with explaining the evolution of species, he proposed that embryos were guided in their development by an *internal mould,* fuelled by *organic molecules,* which recombine into the form of the developing organism. He was anticipating genetics, and even DNA.

Chapter 8

Jean-Baptiste Pierre Antoine de Monet, Chevalier de Lamarck (1744-1829)

Lamarck was a French naturalist, soldier, biologist, and academic, and an early proponent of the idea that biological evolution occurred and proceeded in accordance with natural laws. Text of a lecture given by Lamarck at the Musée National d'Histoire Naturelle, Paris, May 1803:

> *Do we not therefore perceive that by the action of the laws of organisation ... nature has in favourable times, places, and climates multiplied her first germs of animality, given place to developments of their organisations, ... and increased and diversified their organs? Then ... aided by much time and by a slow but constant diversity of circumstances, she has gradually brought about in this respect the state of things which we now observe. How grand is this consideration, and especially how remote is it from all that is generally thought on this subject!*

Lamarck (1744-1829) (Wikipedia)

Lamarck's scientific theories were largely ignored or attacked during his lifetime;

Lamarck never won the acceptance and esteem of his colleagues Buffon and Cuvier, and he died in poverty and obscurity. Today, the name of Lamarck is associated merely with a discredited theory of heredity, namely the *inheritance of acquired traits.*

However, Charles Darwin, Lyell, Haeckel, and other early evolutionists acknowledged him as a great zoologist and as a forerunner of evolution. Charles Darwin wrote in 1861:

> *Lamarck was the first man whose conclusions on the subject excited much attention. This justly celebrated naturalist first published his views in 1801 ... he first did the eminent service of arousing attention to the probability of all changes in the organic, as well as in the inorganic world, being the result of law, and not of miraculous interposition.*

Classical Darwinism has been followed by 'Neo-Darwinism', thanks to our knowledge of DNA, and genes, but what about Lamarckism?

The author must admit that when originally presented with Lamarck's theory of evolution, it had a certain logic. After all, life's experiences one might assume, would have a profound influence on heredity. It was however the absurd over-simplification, by which I mean the giraffe's long neck example, which unfortunately served quickly to attract ridicule. Likewise the strong arms of the blacksmith, or the non-emergence of tail-less mice born to parents who had had the misfortune, in the interests of science, of having their tails amputated.

Jean-Baptiste de Lamarck, half a century before Darwin, expressed the view that certain acquired traits are passed on to future generations. Robyn Lindley in her book *Soma* writes:

> *There is now extant scientific evidence suggesting that nature has evolved a number of molecular mechanisms for us to consciously direct our own evolutionary future. We ignore this at our peril. Why, then, am I writing about Lamarckism, a subject of past and current ridicule? The answer is simple: I am backed by observations that overwhelmingly*

> *support a departure from strict random-mutation/selection as the core of our evolutionary belief system. Yet Richard Dawkins, 'The greatest show on earth. The evidence for evolution', a prime spokesman for neo-Darwinism, has stated that it is an established, unarguable fact that all life on this planet is shaped by Darwinian theory. Although recent decades have seen many supplemental offshoots to this theory, in the end these mechanistic variations reduce to basic neo-Darwinism. Only the Lamarckian concept stands as a truly divergent view of evolution.*

In 1809 Jean-Baptiste de Lamarck published his *Philosophie Zoologique*, which led to his status as a great evolutionist. Though the theory of evolution is popularly attributed to Darwin, it is grossly unfair to Lamarck who was the first to devote an entire book exclusively to the subject. Darwin documented his ideas with exacting observational data unlike Lamarck. And Lamarck's inheritance of acquired traits seems at first to lack the power and generality of Darwin's natural selection. Yet the concept of evolution (including that of humans) was championed by Lamarck and Erasmus Darwin, before Charles Darwin was born.

Examples of inheritance of acquired traits can be taken from Darwin himself. In the *Descent of Man,* Darwin advised young women to learn as much as they could prior to starting families, the expectation being that this would help endow the future children with useful skills. In 1873 Darwin published a note in *Nature*, titled *Inherited Instinct*, which described a dog's violent antipathy toward a butcher who had mistreated it. According to Darwin, the antagonism had been transmitted to at least two generations of the dog's offspring. Darwin wrote that:

> *many special fears, and tastes, which must have been acquired at a remote period, are now strictly inherited.*

Actual experiments testing the mechanism are presented below. The listed experiments and observations, many of them classics in the field, all suggest that a Lamarckian inheritance is operative in Nature. Taken collectively, a large body of experiments offers a neo-Lamarckian view of evolution that is too compelling to ignore. Even human intelligence has Lamarckian ("epigenetic") origins. Now let us examine some interesting research.

In 1927, William McDougall published a paper titled *an experiment for the testing of the hypothesis of Lamarck*. He reported that the first generation of rats made an average of 165 mistakes during their first run through a maze. After some practise the rats learned to carry out the task perfectly. The offspring of these trained rats, however, made an average of only 20 mistakes on their first run as if they had inherited the acquired maze-running ability. It was concluded that Lamarckian transmission was a real process in nature. Others have disagreed and speculated that the effect derives from parent-to-young training. In any event, this is one of those tantalizing experiments that merits more thorough study. It is conceivable, for example, that McDougall's results could be explained by *trail marking* pheromones deposited by the first generation along the favourable route. A second maze for the offspring experiments, identical in design but pheromone-free, could be used to exclude this possibility. The point here is that Lamarckism is a testable concept. The pheromones secreted might be saying 'this is me sensing success', or 'me sensing failure'. Or it might be the quantified accumulation of pheromone on the successful route, compared to less on an unsuccessful route. Or yet again, the presence of pheromones may be devoid of any pathfinder property. In that case, how might the knowledge, present assuredly somewhere in the brain, have an influence on the genes involved in reproduction?

There are numerous examples of a so-called "sire effect" involving heritable genetic transfer. For example, Sobey and Connolly found that when buck (male) rabbits with acquired immunity to the myxomatosis virus were mated with non-immune does, the immunity was passed on to the offspring.

Geneticist T. Sonneborn removed by microsurgery a piece of the cortex (outer surface) of paramecium, a one-celled animal covered by *ciliae* (small hairs) and then reinserted the piece after first rotating it 180° from its original position. It was obvious that the piece had been rotated because the paramecium now had a segment of cilia pointing in the "wrong" direction. Remarkably, the offspring of the paramecium also had an inverted row of cilia. The acquired trait had ostensibly been inherited in true Lamarckian fashion.

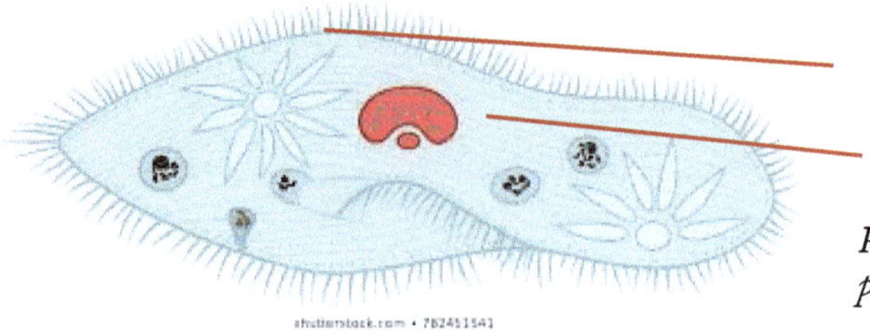

Paramecium (Shutterstock, with permission)

Now let us view some intriguing situations which suggest environment, physical or psychological, can have an apparent hereditary effect.

Exposure of young mice to 2 weeks of an enhanced enrichment program including access to novel toys and elevated levels of social interactions significantly improved their memory and long-term ability to learn. Fascinatingly, the benefits were inherited by their offspring even though the offspring themselves had not been exposed to the enrichment program. Proper controls had been carried out. Thus, offspring of enriched mothers were split into two groups, one of which was given to enriched foster mothers and the other to non-enriched foster mothers. The type of foster mother was found to make no difference. Both groups of offspring profited equally from the stimulating environment experienced by their biological mothers prior to their birth. This seems to be a clear case of heritable adaptation generated by an environmental change.

Hinshelwood showed that some bacteria that had been grown on sublethal levels of toxic drugs survived, and the survivors were then repeatedly transferred to fresh media containing the drugs. They gradually adapted to the drugs to an extent depending on the number of serial passages to which the bacteria had been exposed. After a sufficient number of passages, 100% of the bacteria survived the drugs. These resistant bacteria were then grown for several generations on drug-free media. When the bacteria were transferred to drug-containing media, they all survived, indicating that the original resistance had been maintained during growth and multiplication in the drug-free media. He concluded that he was observing a heritable adaptive change not unlike that proposed by Lamarck. But one assumes, Darwin would have explained this as an example of natural selection.

Hinshelwood's work drew widespread dismissal because his opponents argued that the adaptations had arisen from spontaneous mutations, in a Darwinian-type mechanism, rather than from heritable environmentally induced effects. Indeed, mutations are frequently invoked as an argument against inherited environmental effects even those originating from diet and other factors usually considered non-mutagenic. Although it is difficult to outright eliminate contributions from mutational events, several features of Hinshelwood's data definitely favour a Lamarckian over a neo-Darwinian mechanism:

(a) The doses were sublethal, so this is not a case where all the bacteria are killed off except for a few resistant mutants that continue to multiply.
(b) Drugs of diverse structure all manifested the effect, and none of the drugs is known to be mutagenic.
(c) Rather than an "all or none" behaviour characteristic of many mutations, there was an almost continuous increase in resistance as the number of passages on drug-containing media progressed.
(d) Resistance was expressed more quickly throughout the populations than would be expected from rare mutational events.
(e) Unlike most mutational behaviour, the bacteria regained their original drug-sensitive phenotype after they had been grown for many generations on drug-free media. Gradual reversibility is more adaptive in origin than mutational. Of course antibiotic resistance by the production of beta-lactamase might have been involved.

Another example is that of male mice that had been trained to associate, Pavlov-style, the odour of acetophenone accompanied by mild foot shocks. The offspring of these mice with unexposed females were fearful of the odour although they had never encountered it previously. The fear response was passed on to the next generation even if they were conceived by artificial insemination using sperm of sensitized mice. As is generally true, it is unclear exactly how information is transferred between generations. As with the butcher's dog previously described, and other examples, transmissible behaviour is as yet poorly understood on a genetic basis. Should we be fixated on genes? Could there be another pathway?

Nature itself has provided evidence that heritable transmission of traits arises from effects other than alterations of DNA nucleotide sequences (i.e., mutations). For example, cells in human embryos differentiate into a variety of outwardly expressed appearances (phenotypes) such as nerve, skin, blood, and bone. Since all these cell types in a given human have identical DNA sequences, there must be a form of inheritance that depends on the interaction of the cells with their environments as opposed to classical DNA-based inheritance. The fact that cells of the gut and bone marrow perpetuate themselves for thousands of generations shows that the traits acquired by the differentiated embryonic cells are enduring. However, we now know of the existence of stem cells. These can generate other cells/cell types. They can divide infinitely. This is the so called self-renewal. On the basis of their origin they can be either embryonic, foetal (cord blood) or adult stem cells. Adult stem cells are tissue specific e.g. haematopoietic, i.e. elements of the blood, neural (nerve cell precursors) and muscle stem cells. Tissue specific stem cells are the sources of tissue replacement. Terminally differentiated cells cannot divide, for example red blood cells, however tissue specific stem cells or their derivatives the progenitor/precursor cells can replace them. So tissue specific stem cells are the sources of tissue replacement/regeneration. These stem cells produce precursor cells that differentiate into all the cells of the adult tissues and organs. Adult stem cells or tissue specific stem cells are dormant and will only propagate and produce specialised cells when called into action by a specific signal i.e. trauma, infection, aging. Obviously Lamarck had no idea of the genetics behind the inheritance of acquired traits, just as Darwin had no idea of the genetics behind his survival of the fittest.

Lamarckian inheritance is more recently termed *epigenetic inheritance.* Accordingly, epigenetics is to Lamarckism as neo-Darwinism is to Darwinism. In other words, epigenetics provides a molecular grounding to the inheritance of acquired characteristics. One might well use the term *neo-Lamarckism* instead of *epigenetics* except that the latter carries less emotional and historical baggage with it.

Chapter 9

Georges Cuvier (1769-1832)

Georges Cuvier (Wikipedia)

Jean Léopold Nicolas Frédéric, Baron Cuvier known as Georges Cuvier, was a French naturalist and zoologist, sometimes referred to as the "*founding father of palaeontology*". Cuvier was a major figure in natural sciences research in the early 19th century and was instrumental in establishing the fields of comparative anatomy and palaeontology through his work in comparing living animals with fossils. He is also known for his ideas about the supposed scientific differences between races and for his abuse of Sarah Baartman as part of this 'work'.

Cuvier's work is considered the foundation of vertebrate palaeontology, and he expanded Linnaean taxonomy by grouping classes into *phyla* and incorporating both fossils and living species into the classification. Cuvier is also known for establishing extinction as a fact – at the time, extinction was considered by many of Cuvier's contemporaries to be merely controversial speculation. In his *Essay on the Theory of the Earth* (1813) Cuvier proposed that now-extinct species had been wiped out by periodic catastrophic flooding events. In this way, Cuvier became the most influential proponent of *catastrophism* in geology in the early 19th century.

Among his other accomplishments, Cuvier established that elephant-like bones found in the USA belonged to an extinct animal he later would name as a *mastodon*, and that a large skeleton dug up in Paraguay was of *Megatherium*, a giant, prehistoric ground sloth. He named the pterosaur *Pterodactylus*, and was one of the first people to suggest the earth had been dominated by reptiles, rather than mammals, in prehistoric times.

Cuvier is also remembered for strongly opposing theories of evolution, which at the time (before Darwin's theory) were mainly proposed by Lamarck and Hilaire. Cuvier believed there was no evidence for evolution, but rather evidence for cyclical **creations** and destructions of life forms by global extinction events such as deluges. In 1830, Cuvier and Geoffroy engaged in a famous debate, which is said to exemplify the two major deviations in biological thinking at the time – whether animal structure was due to function or (evolutionary) morphology. Cuvier supported function and rejected Lamarck's thinking.

Cuvier also conducted racial studies which provided part of the *foundation for scientific racism*, and published work on the supposed differences between racial groups' physical properties and mental abilities. Cuvier subjected Sarah Baartman, a Hottentot woman, to examinations alongside other French naturalists during a period in which she was held captive in a state of neglect. Cuvier examined Baartman shortly before her death, and conducted an autopsy following her death that disparagingly compared her physical features to those of monkeys.

Cuvier was critical of theories of evolution, in particular those proposed by his contemporaries Lamarck and Geoffroy Saint-Hilaire, which involved the gradual transmutation of one form into another. He repeatedly emphasized that his extensive experience with fossil material indicated one fossil form does not, as a rule, gradually change into a succeeding, distinct fossil form. A deep-rooted source of his opposition to the gradual transformation of species was his goal of creating an accurate taxonomy based on principles of comparative anatomy. Such a project would become impossible if species were mutable, with no clear boundaries between them. According to the University of California Museum of Palaeontology,

> *Cuvier did not believe in organic evolution, for any change in an organism's anatomy would have rendered it unable to survive. He studied the mummified cats and ibises that Geoffroy had brought back from Napoleon's invasion of Egypt, and showed they were no different from their living counterparts; Cuvier used this to support his claim that life forms did not evolve over time.*

Cuvier's theory on extinction has met opposition from other notable natural scientists like Darwin and Charles Lyell. Unlike Cuvier, they didn't believe that extinction was a sudden process; they believed that like the Earth, animals collectively underwent gradual change as a species. This differed widely from Cuvier's theory, which seemed to propose that animal extinction was catastrophic.

However, Cuvier's theory of extinction is still justified in the case of mass extinctions that have occurred in the last 600 million years, when approximately half of all living species became extinct within a short geological span of 2 million years, due in part to volcanic eruptions, asteroids, and rapid fluctuations in sea levels. At this time, new species rose and others fell, precipitating the arrival of human beings. Cuvier's early work demonstrated conclusively that extinction was indeed a credible natural global process. Cuvier's thinking on extinctions was influenced by his extensive readings of Greek and Latin literature; he gathered every ancient report known in his day relating to discoveries of petrified bones of remarkable size in the Mediterranean region.

Sarah Baartman

Cuvier came to believe that most, if not all, the animal fossils he examined were remains of species that had become extinct. Near the end of his 1796 paper on living and fossil elephants, he said:

> *All of these facts, consistent among themselves, and not opposed by any report, seem to me to prove the existence of a world previous to ours, destroyed by some kind of catastrophe.*

Contrary to many natural scientists' beliefs at the time, Cuvier believed that animal extinction was not a product of anthropogenic (i.e. man-made) causes. Instead, he proposed that humans were around long enough to indirectly maintain the fossilized records of ancient Earth. He also attempted to verify the water catastrophe by analysing records of various cultural backgrounds. Though he found many accounts of the water catastrophe unclear, he did believe that such an event occurred at the brink of human history nonetheless.

This led Cuvier to become an active proponent of the geological school of thought called *catastrophism*, which maintained that many of the geological features of the Earth and the history of life could be explained by catastrophic events that had caused the extinction of many species of animals.

Alongside other French naturalists, Cuvier subjected the aforementioned Sarah Baartman, a South African Khokhoi woman, to exhibition in European freak shows as the "Hottentot Venus". Her existence was really quite miserable and extraordinarily poor. Sarah was literally treated like an animal. In 1815, while Baartman was very ill, Cuvier commissioned a nude painting of her. She died shortly afterwards, aged 26.

Following her death, Cuvier sought out and received permission to dissect her body, focusing on her genitalia, buttocks and skull shape. In his examination, Cuvier concluded that many of Baartman's features more closely resembled the anatomy of a monkey than a human. Her remains were displayed in the *Musee de l'Homme* in Paris until 1970, then were put into storage. Her remains were returned to South Africa in 2002.

Chapter 10

Ernst Haeckel (1834-1919)

Ernst Heinrich Philipp August Haeckel was a German zoologist, naturalist, eugenist, philosopher, physician, professor, marine biologist, and artist who discovered, described and named thousands of new species, mapped a genealogical tree relating all life forms, and coined many terms in biology, including *ecology, phylum, phylogeny* and *Protista*. Haeckel promoted and popularised Charles Darwin's work in Germany and developed the influential but no longer widely held recapitulation theory ("ontogeny recapitulates phylogeny") that claimed that an individual organism's biological development, or ontogeny, parallels and summarises its species' evolutionary development, or phylogeny. In 1866 he met Charles Darwin at Down House in Kent.

Ernst Haeckel (Wikipedia)

Haeckel's affinity for the German Romantic movement coupled with his acceptance of a form of Lamarckism, influenced his political beliefs. Rather than being a strict Darwinian, Haeckel believed that the characteristics of an organism were acquired through interactions with the environment and that ontogeny reflected phylogeny. He saw the social sciences as instances of "applied biology", and that phrase was picked up and used for Nazi propaganda. In 1906

Haeckel founded a group called the *Monist League* to promote his religious and political beliefs.

Haeckel was a flamboyant figure, who sometimes took great, non-scientific leaps from available evidence. For example, at the time when Darwin published *On the Origin of Species by means of Natural Selection (1859),* Haeckel postulated that evidence of human evolution would be found in the Dutch East Indies, now Indonesia. At that time, no remains of human ancestors had yet been identified. He described these theoretical remains in great detail and even named the as-yet unfound species, *Pithecanthropus alalus*, and instructed his students to go and find it.

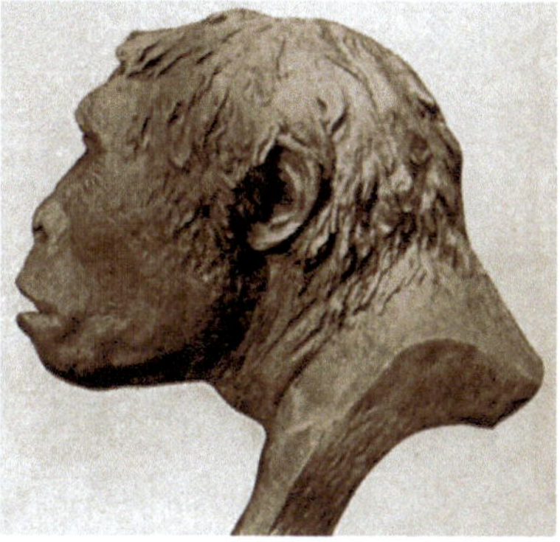
Java Man (Wikipedia)

Consequently the remains of *Java Man* were discovered in 1891, consisting of a skullcap, thighbone, and a few teeth. These remains are among the oldest hominid remains ever found. Java Man was suggested as a potential intermediate form between modern humans and the common ancestor we share with the great apes.

Haeckel also applied the hypothesis of polygenism to the modern diversity of human groups. He became a key figure in social Darwinism and leading proponent of scientific racism, stating for instance:

> *The Caucasian, or Mediterranean man (Homo Mediterraneus), has from time immemorial been placed at the head of all the races of men, as the most highly developed and perfect. It is generally called the Caucasian race, but as, among all the varieties of the species, the Caucasian branch is the least important, we prefer the much more suitable appellation proposed namely, that of Mediterranese. For the most important varieties of this species, which are moreover the most eminent actors in what is called "Universal History", first rose to a flourishing condition on the shores of the Mediterranean... This species alone (with the exception of*

> *the Mongolian) has had an actual history; it alone has attained to that degree of civilisation which seems to raise men above the rest of nature.*
>
> *Haeckel divided human beings into ten races, of which the Caucasian was the highest and the primitives were doomed to extinction. Haeckel claimed that 'the Negro' had stronger and more freely movable toes than any other race, which, he argued, was evidence of their being less evolved, and which led him to compare them to "four-handed" Apes. Haeckel also believed 'Negroes' were savages and that Whites were the most civilised.*

In his *Ontogeny and Phylogeny* Harvard palaeontologist **Stephen Jay Gould** wrote:

> *[Haeckel's] evolutionary racism; his call to the German people for racial purity and unflinching devotion to a 'just' state; his belief that harsh, inexorable laws of evolution ruled human civilization and nature alike, conferring upon favoured races the right to dominate others ... all contributed to the rise of Nazism.*

Haeckel claimed the origin of humanity was to be found in Asia: he believed that the Indian subcontinent was the actual location where the first humans had evolved. He argued that humans were closely related to the primates of Southeast Asia and rejected Darwin's hypothesis of Africa.

Haeckel later claimed that the missing link was to be found on the lost continent of "*Lemuria*" located in the Indian Ocean. He believed that Lemuria was the home of the first humans and that Asia was the home of many of the earliest primates. He thus supported the view that Asia was the cradle of hominid evolution, and claimed that Lemuria connected Asia and Africa, which allowed the migration of humans to the rest of the world.

In Haeckel's book *The History of Creation* (1884) he included migration routes which he thought the first humans had used outside of Lemuria.

When Haeckel was a student in the 1850s he showed great interest in embryology, attending the rather unpopular lectures twice and in his notes sketched the visual aids: textbooks had few illustrations, and large format plates were used to show students how to see the tiny forms under a reflecting microscope, with the translucent tissues seen against a black background. Developmental series were used to show stages within a species, but inconsistent views and stages made it even more difficult to compare different species. It was agreed by many European evolutionists that all vertebrates looked very similar at an early stage, in what was thought of as a common ideal type, but there was a continuing debate from the 1820s between the Romantic recapitulation that human embryos developed through stages of the forms of all the major groups of adult animals, literally manifesting a sequence of organisms on a linear chain of being. Haeckel's aim was a reformed morphology with evolution as the organising principle of a cosmic synthesis unifying science, religion, and art. He was giving successful popular lectures on his ideas to students and townspeople in Jena.

He drew figures for the book, synthesising his views of specimens in Jena and published pictures to represent types. After publication he told a colleague that the images:

> *are completely exact, partly copied from nature, partly assembled from all illustrations of these early stages that have hitherto become known.*

There were various styles of embryological drawings at that time, ranging from more schematic representations to "naturalistic" illustrations of specific specimens. Haeckel believed privately that his figures were both exact and synthetic, and in public asserted that they were schematic like most figures used in teaching. The images were reworked to match in size and orientation, and though displaying Haeckel's own views of essential features, they support von Baer's concept that vertebrate embryos begin similarly and then diverge. Relating different images on a grid conveyed a powerful evolutionary message.

While it has been widely claimed that Haeckel was charged with fraud by five professors and convicted by a university court at Jena, there does not appear to be an independently verifiable source for this claim. Recent analyses (Richardson 1998, Richardson and Keuck 2002) have found that some of the criticisms of Haeckel's embryo drawings were legitimate, but others were unfounded. There were multiple versions of the embryo drawings, and Haeckel rejected the claims of fraud. It was later said that "there is evidence of sleight of hand" on both sides of the feud between Haeckel and Wilhelm His. Robert J. Richards, in a paper published in 2008, has defended the case for Haeckel, shedding doubt against the fraud accusations based on the material used for comparison with what Haeckel could access at the time.

Haeckel was awarded the title of Excellency by Kaiser Wilhelm 2nd in 1907 and the Linnaean Society of London's prestigious Darwin-Wallace Medal in 1908.

Chapter 11

August Weismann (1834-1914)

August Weismann was a German evolutionary biologist, arguably the second most notable evolutionary theorist of the 19th century, after Charles Darwin.

His main contribution involved germ plasm theory, at one time also known as *Weismannism* according to which inheritance (in a multicellular organism) only takes place by means of the germ cells – the gametes such as egg cells and sperm cells. Other cells of the body – somatic cells – do not function as agents of heredity. The effect is one-way: germ cells produce somatic cells and are not affected by anything the somatic cells learn or therefore any ability an individual acquires during its life. Genetic information cannot pass from soma to germ plasm and on to the next generation. Biologists refer to this concept as the *Weismann barrier.* This idea, if true, rules out the inheritance of acquired characteristics as proposed by Lamarck. However, a careful reading of Weismann's work over the span of his entire career shows that he had more nuanced views, insisting, like Darwin, that a variable environment was necessary to cause variation in the hereditary material.

August Friedrich Leopold Weismann FRS (Wikipedia)

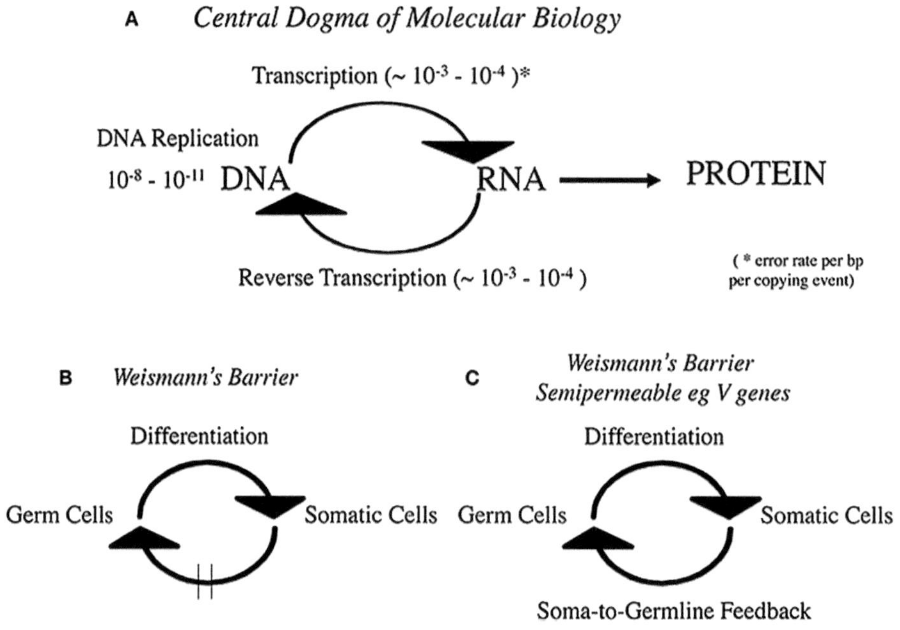

The idea of the *Weismann barrier* is central to the modern synthesis of the early 20th century, though scholars do not express it today in the same terms. In Weismann's opinion the largely random process of mutation, which must occur in the gametes (or stem cells that make them) is the *only* source of change for natural selection to work on. Weismann became one of the first biologists to *deny Lamarckism entirely.* Weismann's ideas preceded the rediscovery of Gregor Mendel's work, and though Weismann was cagey about accepting Mendelism, younger workers soon made the connection.

Even at that time, when the writings of Gregor Mendel on genetics were lying unnoticed, Weismann saw that, since the hereditary substances from the two parents become mixed together in the fertilized egg, *there would be a progressive increase in the amount of hereditary substances unless at some stage there was a compensating reduction.* He therefore predicted that there must be a form of nuclear cell division in which each daughter nucleus receives only half the ancestral germ plasms contained in the original nucleus. This is now known to be the case, the process being the phenomenon of *meiosis.* The cytological work of other investigators proved the correctness of this prediction and enabled Weismann, together with the others, to propose that the germ plasm was located in what were subsequently called the chromosomes of the egg nucleus.

Unlike Darwin, however, Weismann firmly opposed the idea of the inheritance of acquired characters. He put the matter to a practical test in a somewhat naively conceived experiment in which he cut off the tails of mice. With painstaking thoroughness, he observed five generations of progeny of tailless parents, 901 mice in all. Needless to say, they all grew normal tails, and Weismann was able to conclude that mutilations were not inherited.

In later life Weismann established himself as one of the leading biologists of the world, and his lectures on heredity and evolution became famous.

From the first, Weismann was a strong supporter of the theory of evolution by natural selection, as put forward by Charles Darwin and Alfred Wallace. In his 1904 book *The Evolution Theory,* he stated that Darwin's *On the Origin of Species,* when it was published in German in 1859, fell "like a bolt from the blue".

In spite of his enthusiastic support of Darwin, Weismann felt it necessary to disagree with that part of the theory in which Darwin had accepted the Lamarckian view of the inheritance of acquired characters. Weismann disagreed strongly with this concept, both on technical grounds and from experimental evidence (or the lack of it). He first publicly expressed his views on the matter in 1883 in the essay *Heredity*, presented as his inaugural address as prorector of the University of Freiburg. He pointed out the impossibility of a mechanism whereby changes in the external organs and tissues of an animal, induced by environmental stimuli, would be conveyed to the reproductive organs and the germ cells within them and thence to ensuing generations.

Weismann realized it was necessary to suggest some other mechanism for producing the variations necessary for evolution. In this he was not very successful. He spoke as though natural selection could itself act in this way, but he was vague about the details. In *The Germ Plasm* he mentioned "chance nutritive fluctuations" in the germ plasm as giving rise to variations.

The idea for *The Germ Plasm* appears to have stemmed from Weismann's early embryological studies, especially with *Hydra,* where he observed that only certain predetermined cells were capable of giving rise to the germ line and to daughter individuals. He extended the idea to the contents of these cells and proposed that there was a certain substance, or *germ plasm*, which could never be formed anew but only from pre-existing germ plasm. It was transmitted unchanged from generation to generation and controlled all the characters of the individual animals. The idea of the germ plasm seemed (and seems) to some people to be somewhat mystical, as it postulates a completely self-determining substance, which apparently does not obey the laws of the physical world, since it proceeds along a path determined only by itself, unaffected by the surrounding environment.

Weismann made his germ plasm theory all-embracing, in that he attempted to explain not only heredity but also development. In fact, at times he seemed to place greater significance on the latter aspect than on the former and allowed his imagination to get rather out of hand. He proposed that the total hereditary substance of a cell should be called *idioplasm*. Every cell contained idioplasm, while the idioplasm of the germ cells was the germ plasm. The idioplasm was composed of smaller entities called *ids*. Each id in its turn consisted of determinants, each controlling the development of a particular part of the organism. The determinants contained certain groups of *biophors,* the simplest living units, which were thought to consist of *albumen molecules, water and salts.*

Surprisingly, Weismann did not appear to appreciate the full significance of Gregor Mendel's work even after 1900. In *The Evolution Theory,* apropos Mendel's work, Weismann stated:

> *We must postpone the working of this new material into our theory until a very much wider basis of facts has been supplied.*

Yet in some respects Weismann was remarkably foresighted. In his discussion on the "inheritance" of musical ability and other cultural activities, he clearly set out the distinction between biological heredity, based on a transmission of material through the germ cells, and cultural inheritance, resulting from a process of learning of skills and traditions by individuals of each generation from their parents and other individuals in the surrounding society. Again, in regard to the origin of life on Earth, which Weismann discussed in the last chapter of *The Evolution Theory,* he dismissed the possibility that life was brought to the Earth in a meteorite and came out firmly in favour of spontaneous generation – not however of any form of life like that now familiar to us, but of some extremely primitive bodies (*biophors).*

> *These albuminoid substances, could have arisen spontaneously through purely chemico-physical causes, from inorganic materials, under conditions which may no longer exist on the earth. Such views are not far removed from the speculations on the origin of life currently in vogue.*

Chapter 12

Mendel, Genetics and Epigenetics

Unfortunately, none of the authorities quoted in the preceding chapters knew or paid any attention to the work of Gregor Mendel. And this unfortunately included Charles Darwin himself, although they shared this planet for 55 years.

Gregor Mendel is recognised as the father of genetics. He was a monk in Brunn, now Brno. Like many great artists, his work was not appreciated until after his death. He became a monk in 1843 and studied at the University of Vienna. He was at home in the monastery's botanical garden where he spent many hours a day breeding fuchsias and pea plants.

Gregor Mendel (1822-1884) (Wikipedia)

He wanted to create hybrid pea plants. This involved using a paint brush to transfer pollen from the anthers (male) to the pistils (female) of related species. He then observed the results which could take several months. Obviously enormous patience and determination was required. Let us imagine similar experiments were to be performed on different species, see the table.

Species	**Gestation period**
Bacteria e.g. E. Coli in the laboratory	20 minutes
Viruses	12 hours
Peas	70 days
Man	9 months
Elephant	18-22 months

Nowadays, transferable resistance of bacteria to antibiotics requires about two days to detect, and involves genetic transfer by the bacteria. So answers can be obtained within a few days. Mendel's life-form required 70 days to reveal what we now know to be his laws on gene segregation. His observations led to more experiments, which led to unusually prescient conclusions. By simply counting peas bearing the desired characteristics and keeping meticulous notes, Mendel established the principles of inheritance, coined the terms *dominant* and *recessive*, and was the first to use statistical methods to analyse and predict hereditary information. For eight years, Mendel cultivated thousands of pea plants and used a paintbrush to painstakingly transfer pollen from one plant to another to make his crosses (all the while still attending to his duties as a monk and a teacher). In 1865 he presented his work to the Brunn Society for Natural Science and his paper, *Experiments on Plant Hybridisation*, was published the next year. It was over 30 years before his work was appreciated.

Mendel's first law of gene segregation. In this example, the red squares are dominant, the yellow recessive.

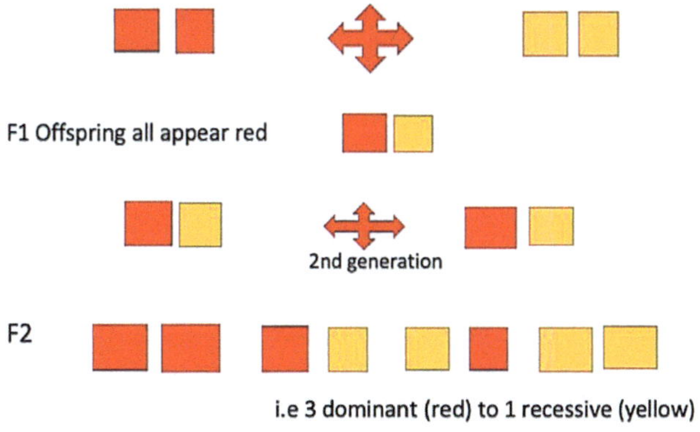

This is known as Mendel's *1st law of gene segregation*. When more than one gene is investigated, they are found to segregate independently. This is known as Mendel's second law.

The genetic constitution is the *genotype*, the actual appearance the *phenotype*.

> *So we see that the concept of creation of species by the construction of genes is within our powers of imagination. All living forms possess proteins. So every living form possesses the genes encoding these proteins. Therefore given the genetic code, theoretically all living forms can have their DNA or RNA artificially produced, and therefore by extrapolation, all forms of life could be produced.*

Meet the deepest fish in the ocean, a new species named by an international team of researchers that discovered it, the **Mariana snailfish**. The Mariana snailfish *(Pseudoliparis swirei)* thrives at depths of up to about 8,000 meters (26,200 feet) along the Mariana Trench near Guam. Its genome has been analysed, and its relation to other species and common ancestors depicted. Elevated hydrostatic pressure can influence gene and protein expression in both one atmosphere-adapted and high pressure-adapted microorganisms. Experiments have documented these effects and described their significance towards understanding the molecular bases of life in deep-sea high-pressure environments.

Snailfish from the Mariana Trench

Does this imply that the environment can, under these circumstances, affect a gene (à la Lamarck)?

Glutamic acid is a functional amino acid that plays important roles in plant nutrition, metabolism, and signal transduction. The rapid and specific induction of transcription factor, kinase/phosphatase and elicitorresponsive genes suggests that glutamate may efficiently amplify its signal and interact with other signalling pathways to regulate metabolism, growth and defence responses in rice. (Lamarck smiles!).

Historians of science have noted that, while Darwin considered the ideas in Wallace's paper to be essentially the same as his own, there were differences. Darwinian evolution as an abstract concept may seem far removed from that attributed to Darwin. When evolution is conceived in most people's minds, it is perceived as involving a progressive increase in cell number and complexity, whereas this need not necessarily be the case. Most organisms have adapted to specific environments. Multicellular life is not necessarily the goal of evolution. The developmental sense of evolution, has traditionally been referred to as the *scala naturae*, popular with science fiction screenwriters. **The next species that evolves from humans may be a herbivorous and apelike animal that has adapted to global warming**. Hopefully, not a virus!

The concept of *common ancestry* leading to evolutionary development is interesting and questionable. Why is it necessary? It only takes us back to the number one, i.e. the totipotent origin from which all phyla, genera, species, and subspecies arose. And over aeons. And before that? When did time start? How? What is time? Is it a construct in our minds to form a framework for events? All sentences require a verb, which invariably incorporates time – past, present, or future. A **black hole** is a region of spacetime exhibiting **gravitational** acceleration so strong that nothing – no particles or even electromagnetic radiation such as light – can escape from it. The theory of general relativity predicts that a sufficiently compact mass can deform spacetime to form a **black hole**. In physics, spacetime is any mathematical model that fuses the three dimensions of space and the one dimension of time into a single four-dimensional continuum. Spacetime diagrams can be used to visualize relativistic effects such as why different observers perceive where and when events occur differently.

What were Erasmus Darwin's *filaments* from which everything arose? The scientific world is pretty uniformly anti-creationist, but it fails, at present, to explain events without the necessity for the incorporation of time. When it does, all may become apparent, the true revelation.

May it be that evolution is like baking a cake? We know about the essential ingredients – baking powder, baking soda, butter, cornstarch, eggs, flour, milk and salt, the oven, switched on, and operating for the required duration. Creationists are under the impression that the oven can do the job a billion times faster than is necessary. Darwin is more realistic, leaving the oven on for aeons.

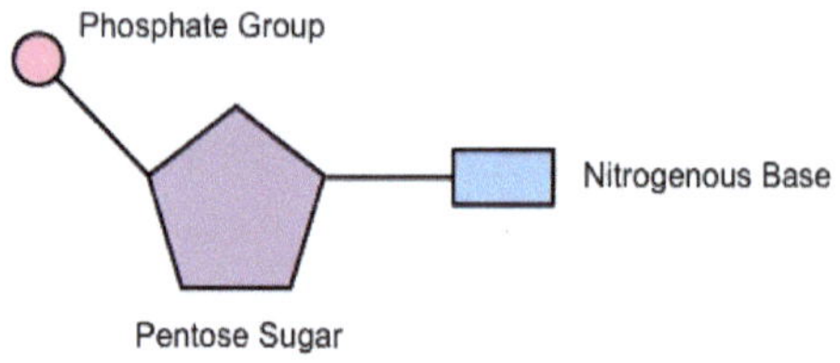

But can genes be created by clever chefs? Indeed they can.

With the discovery of the structure of *deoxyribonucleic acid* (DNA) by **Watson** and **Crick** published in 1953, the basis of the code for heredity was established. Apart from fat and

DNA Structure

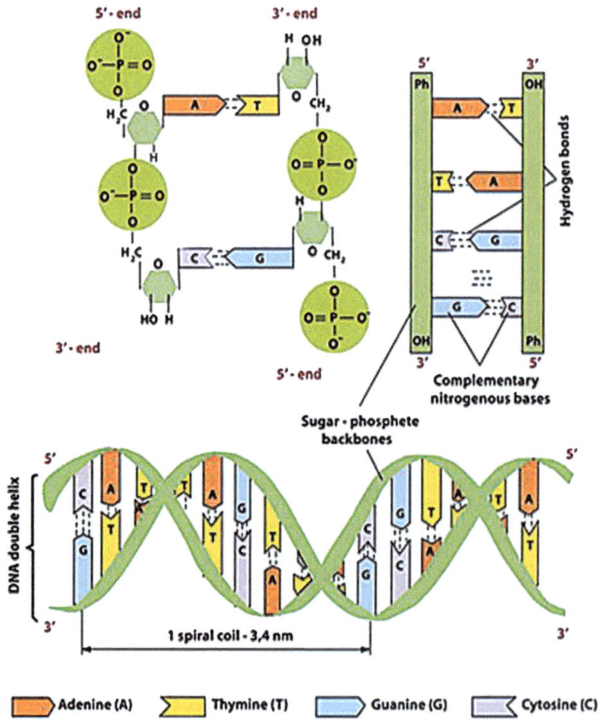

carbohydrate, we are made of protein, which in turn consists of building blocks, called *amino-acids*. Each amino-acid is coded for by a sequence of organic (i.e. carbon containing) chemicals called *nucleotides* which consist of a nitrogen containing base connected to a phosphate (the acidic part) and a sugar (*deoxyribose*). The bases are *adenine* (A) which is always linked to *thymine* (T), *and cytosine* (C) which is always linked to *guanine* (G). Three nucleotides encode an amino-acid. This is called the '**triplet code**'. However in the case of *ribonucleic acid* (RNA), adenine is linked to *uracil* not *thymine*.

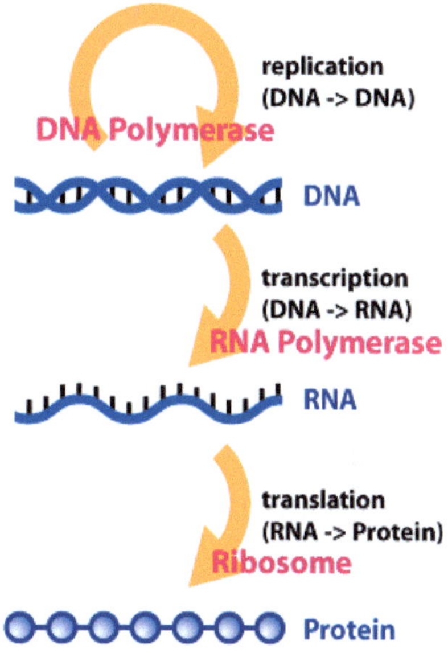

DNA is transcribed into RNA, (the R refers to ribonucleic acid), which is slightly different to deoxyribonucleic acid, and in RNA, as stated, thymine is replaced by *uracil*. Polymerases are enzymes (proteins) which catalyse reactions.

Perhaps RNA was the first nucleic acid to exist and DNA evolved from it, such that the changes in the sugar, ribose for deoxyribose, and *uracil* for thymine might have some advantage. Possibly, it is less susceptible to degradation by radiation, but this is not certain.

What about so-called 'essential amino acids'? As a medical student, I was taught that most of the required amino acids could be synthesised by humans, but not all. The latter were termed 'essential amino acids', and had to be obtained by the diet. The mnemonic used to remember them was:

> AV HILL MP TT. This stands for: arginine, valine, histidine, isoleucine, leucine, lysine, methionine, phenylalanine, threonine and tryptophan.

It seems plants possess the means of producing them, whereas we do not.

Amino acids might become essential for a given organism because the gene(s) required are lost. How? If there is a mutation, and the message for that particular amino acid is lost, but the amino acid can be found in the diet, the organism may thrive despite its genetic loss. Environmental constraints under which organisms evolve are dynamic over time. If an amino acid is available for uptake, the selective pressure to keep intact the genes responsible for that pathway might be lowered, because they would not be constantly expressing these biosynthetic genes. Without the selective pressure, the biosynthetic routes might be lost or the gene could allow mutations that would lead to a diversification of the enzyme's function. Following this logic, amino acids that are essential for certain organisms might not be essential for other organisms subjected to different selection pressures. Could this be interpreted as an example of Lamarckism, in that the entry of an essential amino acid (by diet), i.e. the environment, could affect the DNA of the host (Man), so that the species though unable to make this, and about nine other amino acids would desist from developing the necessary DNA, and concentrate on other amino-acids? Viva Lamarck! By what mechanism would the natural selection of those genes not expressed lead to their removal?

Can changes in the heritable phenotype not involving alterations in the DNA sequence actually occur?

Epigenetics is the study of features that are "on top of" or "in addition to" the traditional genetic basis for inheritance.

The organisation of the host's genetic library held within the cell's nucleus consists of DNA that is wound around spools, called **histones**. The spools influence the DNA such that they shuffle the DNA into structural units called **nucleosomes**. These become bundled into **chromatin** fibres, the totality of which constitute **chromosomes**.

Nucleosomes possess tails. It may be, i.e. it has been hypothesised, that the reading (transcription) of the DNA code is effected by the chemical modifications to the histones, which consequently recruit other proteins to their respective domains which may then alter chromatin structure or promote transcription.

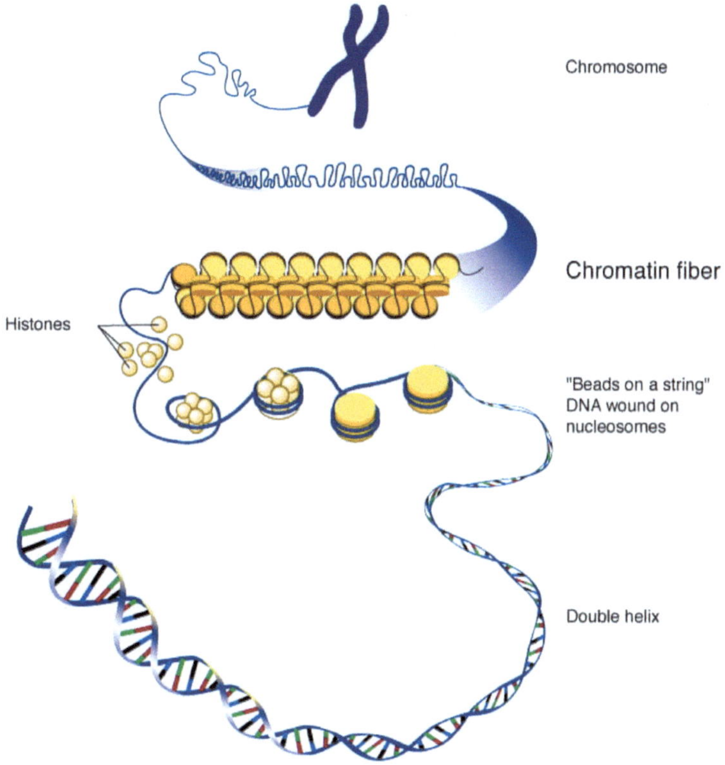

Chromatin

Together with similar modifications such as *DNA methylation* it is part of the **epigenetic code**.

When more than one theory of anything is postulated, history suggests that one may be right, and by inference, the other wrong. In the 19th century, we had the theory that only antibodies protected against infections, as propounded by **Behring** and **Kitasato**. The rival theory was proposed by **Eli Metchnikoff**, namely that protection against infections was due to cells which could engulf and kill germs, a process termed **phagocytosis**. Subsequently it has been shown that both are correct, in fact, antibodies can attach to germs by a figurative arm (the 'fab' fragment, fab meaning antigen binding – see diagram) and hold out a leg that is seized by the phagocytic cell (the Fc portion, 'fragment crystallizable' so called because it can be crystallized – see diagram overleaf).

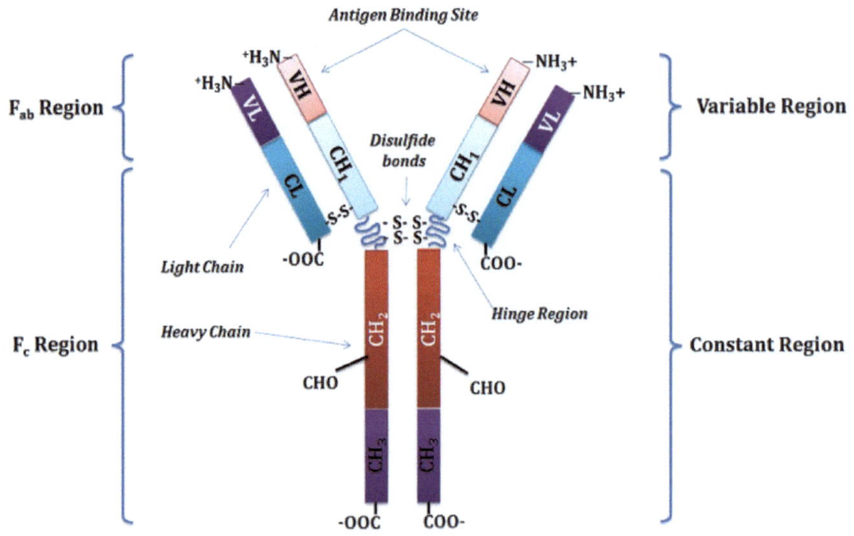

Variable (V) and Constant (C) Regions of an Antibody

Epigenetics: The Science of Change

Bob Weinhold (2006). This is reproduced from 'Environmental Health Perspectives 2006 (March) 114(3): A160-167'

> For nearly a century after the term "epigenetics" first surfaced on the printed page, researchers, physicians, and others poked around in the dark crevices of the gene, trying to untangle the clues that suggested gene function could be altered by more than just changes in sequence. Today, a wide variety of illnesses, behaviours, and other health indicators already have some level of evidence linking them with epigenetic mechanisms, including cancers of almost all types, cognitive dysfunction, and respiratory, cardiovascular, reproductive, autoimmune, and neurobehavioral illnesses. Known or suspected drivers behind epigenetic processes include many agents, including heavy metals, pesticides, diesel exhaust, tobacco smoke, polycyclic aromatic hydrocarbons, hormones, radioactivity, viruses, bacteria, and basic nutrients.

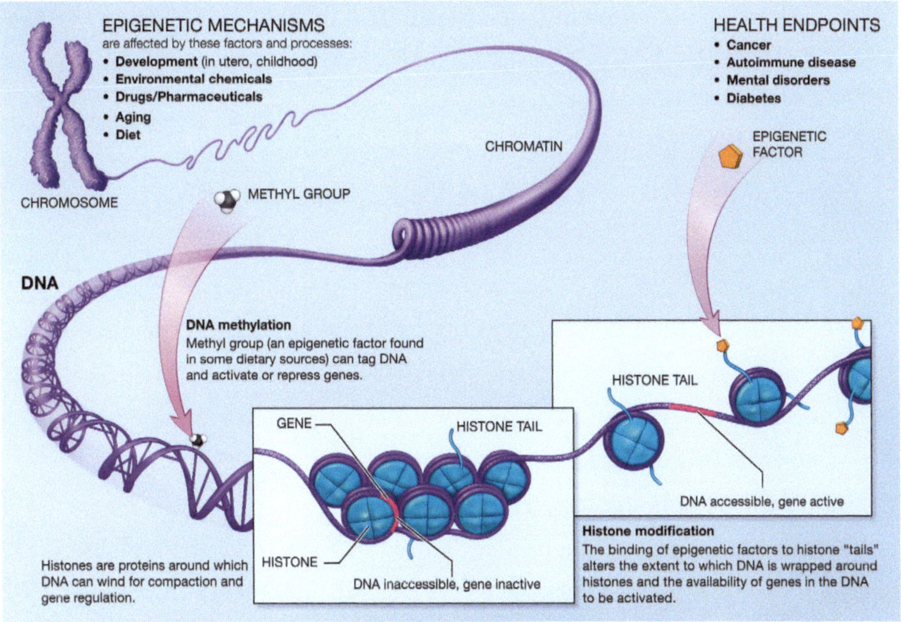

In the past five years, and especially in the past year or two, several groundbreaking studies have focused fresh attention on epigenetics. Interest has been enhanced as it has become clear that understanding epigenetics and epigenomics—the genomewide distribution of epigenetic changes—will be essential in work related to many other topics requiring a thorough understanding of all aspects of genetics, such as stem cells, cloning, aging, synthetic biology, species conservation, evolution, and agriculture.

Multiple Mechanisms

The word "epigenetic" literally means "in addition to changes in genetic sequence." The term has evolved to include any process that alters gene activity without changing the DNA sequence, and leads to modifications that can be transmitted to daughter cells (although experiments show that some epigenetic changes can be reversed). There likely will continue to be debate over exactly what the term means and what it covers.

The following is reproduced from the below-mentioned article FRONTIERS IN LIFE SCIENCE, 2016 VOL. 9, NO. 4, 347–370

'Epigenetics: from the past to the present Nubia Andrea Villota-Salazara, Artemio Mendoza-Mendozab and Juan Manuel González-Prieto'

Epigenetics is an increasingly important field because it provides a broader outlook of biological systems. Nowadays, many biological phenomena are approached from the epigenetics perspective; however, this area has not been so well known. Despite that the concept of epigenetics firstly appeared in 1939, only in recent years the term has been popularly used in the scientific literature. The idea of epigenetics of C. Waddington related the causal elements of the development and the embryology for explaining that in the development of organisms, both preformed and no preformed elements are involved. After the report of genetically identical cells with different phenotypes, the detailed studies on gene regulation of proteins, the existence of a genetic program of development in organisms, the evidence of no genetic transmission of traits and the molecular mechanisms of DNA organisation in the nucleus; the current concept of epigenetics began, but even today, it is still under intense debate. Currently, we know that in eukaryotes the occurrence of epigenetic phenomena is related to the structure of chromatin and its configurations. DNA is organized into chromatin within the nucleus. Chromatin is a complex with DNA, histone proteins and no histone proteins, which have two functional states, heterochromatin and euchromatin. The nucleosome is the first level of chromatin organisation that enables DNA packaging, regulating gene expression and making possible epigenetic phenomena. Chromatin allows different configurations of the same genome, causing various epigenomes and then different phenotypes. Histone modifications, DNA methylation and the

> activity of nonncRNA, are the major mechanism of epigenetic regulation in eukaryotes. Epigenome represents the connection between environment and genome. Environmental influences can affect the programming of the genome, activating specific pathways in cells that, through the mechanisms of epigenetic regulation, promote the stable remodeling of chromatin, changing gene expression and phenotype. There is evidence that some of these changes can be inherited to the progeny even when the environmental trigger is over. In humans, environmentally induced modification in the epigenome has also been linked to a variety of pathologies, nevertheless, DNA and chromatin alterations are reversible and the implementation of epigenetic therapies for the treatment of epigenetically based diseases is an area of ongoing research and shows amazing perspectives. Undoubtedly, a major breakthrough could be expected. Parallel to the report of more reference epigenomes, the development of new technologies, laboratory techniques and tools for management and integration of information obtained, it will be possible to detail the interactions among environment, epigenome and genome and understand the role of these interactions in the inheritance of traits, adaptation and evolution of organisms.

Interestingly, there are enzymes called *reverse transcriptases,* used by certain viruses, including the human immune-deficiency virus (HIV a cause of AIDS), which are able to do this, as a consequence of which the viral RNA gets into the host cell's DNA. There are other examples of eukaryotic DNA being reverse transcribed. Reverse transcriptase enzymes are involved in genetic diversity and in the process of ageing in eukaryotic cells. We might imagine the world's motor-bikes, nearly all of which can only go forwards, because by and large, motorbikes don't have a reverse gear. (Except that there are some Harleys which do!) Likewise, Mother Nature can, in circumstances not yet fully understood, go into genetic reverse, as a consequence of which the environment can directly modify the host's DNA, a phenomenon utilising *reverse transcriptase* which is an enzyme used to generate complementary DNA from an RNA template, a process termed *reverse transcription.*

Inheritance is a term understood by people as applied to features such as skin colour, tendency to baldness, stature, and various others. It has always been to the fore with animal breeders, though in-detail knowledge was absent prior to the recognition of genes (Gregor Mendel), and DNA (Watson and Crick). One problem with the basic concept of DNA (by analogy to a brick works), is that the possession of thousands of bricks in a pile doesn't explain how they are arranged in order to make a building. Whilst it can readily be shown to encode proteins, it is more difficult to expand this into a scheme for encoding different structures, e.g. liver cells versus lung cells et seq. Both contain the whole gamut of DNA. Charles Darwin formulated his hypothesis of *Pangenesis* with its hereditary molecules, termed *gemmules,* that are thrown off by all cells of the body and assembled in the reproductive cells. However there is no evidence as to what a gemmule looks like or of what it consists. However the discovery of circulating nucleic acids, including small ribonucleic acids (RNAs) and prions (which are misfolded proteins with the ability to transmit their misfolded shape onto normal variants of the same protein, and can cause various diseases, e.g. mad cow), provide striking evidence in favour of the chemical existence of these hypothetical gemmules. There is also a body of experimental evidence for the inheritance of acquired characteristics, graft hybridisation, and other "doubtful" phenomena. The rediscovery of Darwin's Pangenesis, if and when it is admitted, would, like the rediscovery of Mendel's work, have a tremendous impact on genetics, evolution, cell biology, and the history of science.

According to Darwin's Pangenesis, a reserve of gemmules in the cells and tissues of the rest of the body serves for regeneration in case parts are lost. For example, if the leg of a salamander has been removed, the leg gemmules, which are present in the body, can migrate to the cut surface, unite with the cells of the cut surface and develop into a new limb, identical to the old one.

In the history of biology, neglecting certain discoveries is not uncommon. It is well known that Mendel's experiment on plant hybridisation was ignored for over 150 years. However, Darwin's idea that pangenetic gemmules are the molecular carriers of hereditary characters and that they are diffused through the tissues or from cell to cell may in fact be the case. It is supported by the discovery of circulating nucleic acids in human blood and plant sap and the results of experimental work in inducing hereditary changes through blood transfusion in animals and through grafting in plants.

According to Darwin's Pangenesis, the appearance of the same characters in a parent and its offspring was made possible by the production of gemmules by all cells of the parent's body. It is of interest to note that a previous mating may have an influence on a female's offspring by a subsequent mating. This is termed *Telegony*, and is manifest *through the diffusion, retention, and action of the gemmules included within the spermatozoa of the previous male.* In order to explain development, Darwin supposed that differentiation was a function of the gemmules: *The organic units, during each stage of development throw off gemmules, which, multiplying, are transmitted to the offspring. In the offspring, as soon as any particular cell or unit in the proper order of development becomes partially developed, it unites with (or, to speak metaphorically, is fertilized by) the gemmules of the next succeeding cell, and so onwards.* Through his Pangenesis, Darwin tried to account for a considerable variety of biological phenomena and to unite them under a common theory.

Can blood transfusions influence recipients' DNA and by this means, affect future generations? Soon after Darwin's pangenetic theory was published, **Francis Galton** (1822-1911) designed a series of blood transfusion experiments on differently pigmented rabbits to test its validity. He found no evidence in support of the existence of Darwin's gemmules and the concept of Pangenesis was largely abandoned. But in the early 1950s, **Sopikov** demonstrated that repeated injections of the blood of an Australian breed of rooster with black feathers into white leghorn hens and the subsequent mating of these hens with roosters of the same breed (white leghorn) yielded progeny of a modified inheritance. Does the blood carry Darwin's supposed gemmules? The discovery of extracellular nucleic acids in the circulation was reported by **Mandel** and **Métais** in 1948 providing direct evidence for the chemical existence of Darwin's hypothetical gemmules. These investigators observed the presence of circulating DNA and RNA in the plasma of healthy and sick individuals.

This work was particularly remarkable as it was only a few years following the demonstration that DNA is the material of inheritance, and this report even preceded the classical paper on the double-helical structure of deoxyribonucleic acid (DNA) by **Watson** and **Crick** (1953). It was unfortunate, therefore, to observe that there was essentially no interest in circulating nucleic acids in the subsequent decade following this pioneering work.

Darwin (1875) imagined that gemmules were *inconceivably minute and numerous as the stars in heaven* and that *many thousand gemmules must be thrown off from the various parts of the body at each stage of development.* Today, we know that small ribonucleic acid (RNA) molecules, particularly microRNAs, can be secreted from mammalian cells and circulate in blood and other body fluids. Maybe one day, if there are any gemmules capable of repairing transected nerve cells (neurons), we may be able to effectively treat those unfortunate patients who have suffered spinal cord injury.

Could the environment affect the variation attributed to heredity? Darwin believed this to be the case. He clearly stated:

> *There can be no doubt that the evil effects of the long-continued exposure of the parent to injurious conditions are sometimes transmitted to the offspring.*

In a letter to *Nature*, he claimed that fear in animals such as predator avoidance could be strictly inherited. The famous Russian physiologist, **Ivan Pavlov**, accepted the possibility of the inheritance of acquired habit and believed that some of the conditioned newly formed reflexes eventually became transformed into unconditioned ones through heredity.

Later, he reported the results of his experiments, showing that the conditioned reflexes – that is, the highest nervous activity – were inherited (Pavlov, 1923).

> *The latest experiments (which are not yet finished) show that the conditioned reflexes, i.e., the highest nervous activity, are inherited. At present some experiments on white mice have been completed. Conditioned reflexes to electric bells are formed, so that the animals are trained to run to their feeding place on the ringing of the bell. The following results have been obtained: The first generation of white mice required 300 lessons. Three hundred times it was necessary to combine the feeding of the mice with the ringing of the bell in order to accustom them to run to the feeding place on hearing the bell ring. The second generation required, for the same result, only 100 lessons. The third generation learned to do it after 30 lessons. The fourth generation required only 10 lessons. The last generation which I saw before leaving Petrograd learned the lesson after 5 repetitions. The sixth generation will be tested after my return. I think.*

Pavlov later doubted this research, and he essentially retracted it. The reason is unclear, but may have to do with reasonings that were lost in translation.

McDougall (1938) demonstrated that, when rats were forced by unpleasant experience to learn the way to a food source, they gave birth to offspring that learned and acquired the habit more readily than did their parents and this continued with successive generations.

There are many examples, now often included under the blanket term *transgenerational epigenetic inheritance*, of altered environmental conditions resulting in heritable changes in gene expression that occur without a change in DNA.

Does the blood carry Darwin's supposed gemmules? Lo and colleagues (1997) demonstrated that fetal DNA was present in peripheral maternal blood, shedding new light on *telegony*. Over the past several decades, a growing body of evidence has accumulated indicating that nucleic acids are released and circulate in the blood, either because of apoptosis (programmed cell death) or necrosis (death of a cell by other cause e.g. infection, or insufficient blood supply), or because of a new mechanism of active release (Gahan 2013). In recent years, thousands of different RNAs have been identified in mammalian sperm, which supports Darwin's idea that *almost infinitely numerous and minute gemmules are contained within each bud, ovule, spermatozoon, and pollen grain.* Most recently, Gapp and colleagues (2014) demonstrated that stress in early life alters the production of microRNAs in the sperm of mice, which results in depressive behaviours in subsequent generations. Szyf (2014) proposed that microRNAs derived from the brains of mice that had undergone stressful experiences could make their way into the reproductive organ through the circulatory system and could then target the specific gene in sperm.

Charles Darwin in his book, *On the Origin of Species* wrote in regard to animal breeding:

> *We shall best understand the probable course of natural selection by taking the case of a country undergoing some physical change, for instance of climate. The proportional numbers of its inhabitants would almost immediately undergo a change, and some species might become extinct. We may conclude, from what we have seen of the intimate and complex manner in which the inhabitants of each country are bound together, that any change in the numerical proportions of some of the inhabitants, independently of the change of climate itself, would most seriously affect many of the others. If the country were open on its borders, new forms would certainly immigrate, and this also would seriously disturb the relations of some of its former inhabitants. Let it be remembered how powerful the influence of a single introduced tree or mammal has been shown to be. But in the case of an island, or of a country partly surrounded by barriers, into which new and better-adapted forms could not freely enter, we should then have places in the economy of nature which would assuredly be better filled up, if some of the original inhabitants were in some manner modified; for, had the area been open to immigration, these*

same places would have been seized on by intruders. In such case, every slight modification, which in the course of ages chanced to arise, and in any way favoured the individuals of any of the species, by better adapting them to their altered conditions, would tend to be preserved; and natural selection would thus give free scope for the work of improvement.

Nor ought we to think that the occasional destruction of an animal of any particular colour would produce little effect: we should remember how essential it is in a flock of white sheep to destroy every lamb with the faintest trace of black.

But nowadays, paradoxically, black sheep are more valuable than white. Today, the wool from black sheep is in great demand from the fashion industry...

We got the term black sheep of the family because for hundreds of years when a black lamb was born, they were sold for meat because the commercial market couldn't dye black wool. Evolutionary biology is the subfield of biology that studies the evolutionary processes that produced the diversity of life on Earth, starting from a (presumed) single common ancestor. These processes include natural selection, common descent, and speciation.

Jean-Baptiste Lamarck in his 1809 book *Philosophie Zoologique* was the first to describe a fully formed scientific theory of evolution which included the fantastic concept of one *species* giving rise, over time, to another. The term then used was *transmutation of species*, which is an archaic term for evolution.

What is evolution? According to the Oxford Dictionary:

> *The idea of organic evolution was proposed by some ancient Greek thinkers but was long rejected in Europe as contrary to the literal interpretation of the Bible. Lamarck proposed a theory that organisms became transformed by their effort to respond to the demands of their environment. Lyell demonstrated that geological deposits were the cumulative product of slow processes over vast ages. This helped Darwin towards a theory of gradual evolution over a long period by the natural selection of those varieties of an organism slightly better adapted to the environment and hence more likely to produce descendants. Combined with the later discoveries of the cellular and molecular basis of genetics, Darwin's theory of evolution has, with some modification, become the dominant unifying concept of modern biology.*

Darwin's basic tenet was that all different species came from common ancestors, though he never said how many, or if there was only one, what it was. Wallace thought likewise, in fact they published jointly to this effect in 1858. Central to both authors was the observation of natural selection. Selection would be by survival and reproductive capacity, though the latter could be influenced by choice of mate. But the evolution of another species from common ancestors is far more contentious. The fact that an occasional fossil is discovered exhibiting features common to different lines of species doesn't prove that this is the actual evolutionary pathway. If at some time in the future, someone digs up an amphibious car, or a car equipped with wings, would they conclude that boats and aeroplanes were derived from a common ancestor, to wit, the car?

A Daf amphibious car. (Courtesy of the Daily Telegraph*) An amphibian clearly evolved from the land-based four-wheeled contraption known as a motor car. Thus far not flourishing*

An aeroplane, clearly evolved from the motor-car, could yet evolve into a practical device when all the roads are clogged up

A fossil of a feathered dinosaur, 'clearly evolved from a land-based creature, but also capable of flight' (Wikipedia)

Darwinism is a theory of biological evolution developed by the English naturalist Charles Darwin and others, stating that all species of organisms arise and develop through the natural selection of small, inherited variations that increase the individual's ability to compete, survive, and reproduce.

Neo-Darwinism is generally used to describe any integration of Darwin's theory of evolution by natural selection with **Gregor Mendel's** theory of genetics.

Evolution was already subconsciously recognised by farmers and fanciers as the consequence of their (unnatural) selection i.e. breeding in order to develop the characteristics which they wanted. Darwin realised that Nature practised the same, except that survival and reproduction was the driving force, not Man's wishes. He wrote:

> *On the Origin of Species by Means of Natural Selection, or the Preservation of Favoured Races in the Struggle for Life.*

He also wrote about orchids, insects, barnacles and corals. One of Darwin's disadvantages, as viewed nowadays, was that he knew nothing of the mechanisms involved in genetics. Mendel's discoveries passed him by. This could not have happened in today's world, with the benefit of information technology.

The evolutionary process by which two related populations diverge into separate species is called **speciation**.

The degree of divergence between geographically separated populations will gradually increase the longer they are separated. As the genetic divergence between the related populations increase, their genetic compatibility consequently decreases. Eventually, the two populations will diverge to an extent where they can no longer interbreed if returned to a shared environment. When two populations can no longer interbreed and produce fertile, viable offspring they are considered to be a separate species.

Darwin was clearly intrigued by the fossils and wildlife he encountered on his voyage. At the time however, he was preoccupied by his geological interest. In 1858, he received from **Alfred Wallace** an essay which amazingly had a lot in common with his theory. They agreed to publish jointly the concept of evolutionary descent involving survival advantage as the explanation for diversity. Darwin realised the provocation potential of including human origins which Wallace had suggested, and advised against it, citing prejudice i.e. religious. In June 1858, Wallace wrote to Darwin describing natural selection. Darwin was stung into action, his very ideas on natural selection apparently forestalled, but eventually they agreed for a joint presentation to the **Linnaean Society** scheduled for 1st July. Tragically, Darwin's son died of scarlet fever on 28th June, so Darwin did not attend. The title of the presentation was *On the tendency of Species to form Varieties, and on the Perpetuation of Varieties and Species by Natural Means of Selection.*

Whereas one might have anticipated considerable impact of the theory, the president of the Linnaean Society failed to grasp its significance, stating that nothing of significance had come to light during that particular year! In the event, *On the Origin of Species* went on sale in November 1859 and the entire stock of 1,250 copies sold out rapidly. Darwin made the case for common descent pointing out similarities between humans and other mammals, and the perceptive conclusion that if all offspring survived birth, there would be insufficient means for them all to survive, unless they possessed some advantage, however small, in other words, *natural selection*. And consequently, those favoured by survival would propagate, thus maintaining their advantage. He concluded:

> *There is grandeur in this view of life, with its several powers, having been originally breathed into a few forms or into one; and that, whilst this planet has gone cycling on according to the fixed law of gravity, from so simple a beginning endless forms most beautiful and wonderful have been, and are being, evolved.*

These words display a magnificent use of prose and contain, with amazing economy, the place on the stage of the universe, of the observed facts in regard to evolution, without overtly stepping into the maelstrom of creation.

Wallace's take on natural selection was as follows:

> *It then occurred to me that these causes or their equivalents are continually acting in the case of animals also; and as animals usually breed much more quickly than does mankind, the destruction every year from these causes must be enormous in order to keep down the numbers of each species, since evidently they do not increase regularly from year to year, as otherwise the world would long ago have been crowded with those that breed most quickly. Vaguely thinking over the enormous and constant destruction which this implied, it occurred to me to ask the question, why do some die and some live? And the answer was clearly, on the whole the best fitted live ... and considering the amount of individual variation that my experience as a collector had shown me to exist, then it followed that all the changes necessary for the adaptation of the species to the changing conditions would be brought about ... In this way every part of an animal's organisation could be modified exactly as required, and in the very process*

> *of this modification the unmodified would die out, and thus the definite characters and the clear isolation of each new species would be explained.*

It should be stated forthwith that epigenetics is not in direct conflict with natural selection. Both models invoke the idea that favourable traits (whether acquired or mutational in origin) can be passed on preferentially to the offspring, thereby perpetuating the trait ("natural selection"). But epigenetic inheritance does provide an additional source of variation, *derived from environmental conditions*, which is not included in neo-Darwinian theory and the many current off-shoots based on it. The difference between the two constructs is critical. One creates change in response to external stimuli, the other creates change according to random alterations in DNA sequencing. Now to the molecular aspects of transgenerational epigenetic inheritance.

The term *epigenetics* (meaning "outside conventional genetics") refers to the generation of heritable variations that originate purely from **environmental** effects, whereas classical mutations in DNA sequences play no role. Thus, an environmentally induced phenotype can transmit its respective phenotype to its descendants even when the stimulus that originally triggered the phenotypical variation is no longer present. As alluded to above, this occurs when, for example, a human's fibroblast divides in culture to give a fibroblast, and a keratinocyte divides in culture to a give a keratinocyte. As an example we have a liver cell, possessing identical DNA to a skin cell, yet when dividing to produce a next generation cell, produces a cell of the appropriate type i.e. another liver cell.

Perhaps the most straightforward mechanism for inheritance of an acquired trait involves a positive feedback model. Imagine an environmental perturbation that "turns on" a previously silent gene, thereby producing (directly or indirectly) a transcription product that preserves the newly created activity of that same gene. If even low levels of the regulatory product are passed on to daughter cells, the activity of the gene will be inherited. Examples of such self-sustaining regulatory loops acting at the transcriptional and post-transcriptions levels are known. Such behaviour must be classified as an epigenetic or neo-Lamarckian adaptation in that an acquired trait has been passed on from generation to generation *in the absence of any mutational modifications of DNA sequences.* Other transgenerational epigenetic mechanisms have been proposed including noncoding RNA, specific DNA methylation, and chromatin effects.

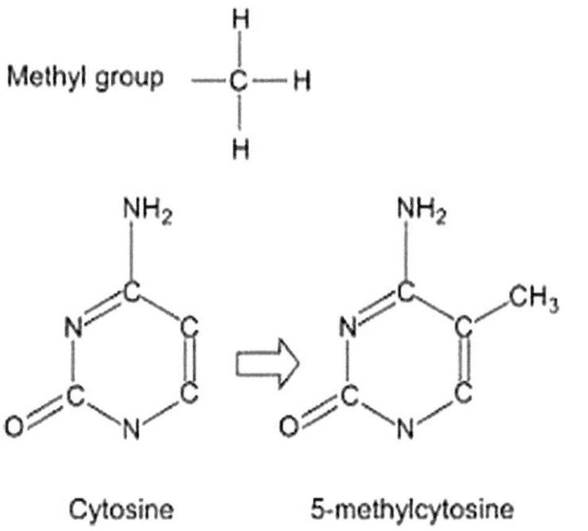

DNA methylation involves an enzyme-catalysed insertion of a methyl group into cytosine, as depicted below or, less commonly, into adenine (two of the four DNA bases).

Methylation can protect a cell that has been virally infected, and which as a consequence would otherwise be beholden to the virus' DNA instruction, by switching the genes into the 'off' position. In other words, epigenetic signalling can switch a gene off. But what if the gene is a tumour-suppressing one? Under these circumstances abnormal hypermethylation could lead to certain cancers owing to the silencing of the afore-said tumour-suppressor genes. DNA methylation directs cell differentiation during embryonic development as well as the ability of these differentiated cells to reproduce with high fidelity. It is at this point that a methyl is introduced into the new G•C unit to give a G•MeC pair, and in this manner the methylation pattern of the original DNA strand is maintained. The important point here is that this mechanism of information transfer occurs independently from an actual sequence mutation. Modification of the DNA sequence occurs only after the DNA molecule has been synthesized. If some sort of environmental perturbation causes a change in the rate or location of methylation, a new inherited phenotype is possible. *Lamarck's theory is, in summary, founded on chemistry.* Note that a significant number of genes in sperm are known to escape epigenetic reprogramming, which scrubs the DNA free of methylation.

As previously described, chromosomes are not merely strands of DNA. Chromosomes are an elaborate complex of DNA, protein, and RNA called chromatin. Thus, DNA is wrapped around a core of so-called **histone proteins** to form **nucleosomes**, the basic subunit of all chromatin. In addition there are a great variety of non-histone proteins that bind directly to the DNA and that include enzymes (e.g. RNA polymerase) as well as proteins that control higher-order structure (e.g. organisation within the chromosomes).

DNA-associated proteins affect the maintenance of gene activity and can be carried from one cell generation to the next. Quite how inheritance of the appropriate state of DNA activity by daughter cells is achieved isn't as yet understood. We can imagine, however, that environmental changes can induce, and in some cases direct, changes in the copying system. Epigenetic variants work faster to achieve adaptation than those accompanying random and generally harmful DNA mutations. Evidence for this is that whereas a given mutation is seldom reproduced concurrently in multiple cells or organisms, this is not true of epigenetic variations induced by environmental changes. The speed and efficiency of epigenetics would appear to support Lamarckian inheritance. However it is quite possible that new mechanisms of epigenetic information transfer will be discovered in the future.

The most important question in science, "What is the evidence?" must, of course, be applied to the cranial feedback theory of human evolution. Although hard evidence is as yet lacking, there are numerous observations pointing directly to its plausibility. Some of these are now cited.

It has been shown that IQ worldwide has increased substantially in recent years, a phenomenon referred to as the Flynn effect. Inasmuch as IQ measures intelligence, people are getting measurably smarter in recent decades owing in part, one surmises, to improved and more widespread education.

Genetic changes from trauma suffered by Holocaust survivors have been demonstrated to be passed on to their children. It appears therefore that life experiences of humans, not just their genes, may affect subsequent generations.

British and Swedish researchers studied carefully documented historical records from Overkalix, an isolated community in northern Sweden. They found that the paternal grandfather's food supply (i.e. whether or not there was a famine during their pre-puberty period) was closely linked to the mortality of the grandsons (but not the granddaughters). It was concluded that the environment might be able to modify the germline. In other words, exposure of a group of men to a particular diet had a profound effect on the viability (disease resistance etc) of men living two generations downstream. Since diet is normally considered an environmental factor, inheritance of the viability trait must be Lamarckian (i.e. acquired) rather than the result of diet-induced neo-Darwinian style mutations. It is not clear at this point for how many generations the diet effect persists.

Cambodian people traumatized during the Khmer Rouge genocide tended to have children with depression and anxiety. Children of Vietnam war veterans in Australia have higher suicide rates than the general population. Obviously, the trauma's effect derives in part from social factors including how the parents might have treated their children. As already stated, a recent study concluded that stress in early life alters the production of small RNAs (microRNAs) in the sperm of small mammals, causing depressive behaviours and metabolic glitches that persist in their progeny. Likewise, human sperm also experiences subtle environmentally-based changes in microRNAs, resulting in a plethora of other effects.

A possible intriguing example is the Dutch famine of World War II. Starving mothers who gave birth during the famine had children who were more prone to obesity and other metabolic disorders – and so were their grandchildren.

If diet, trauma and obesity can impact our genes, is it far-fetched to postulate that brain chemistry can do likewise?

Will the human race, averaged over the entire population, become smarter and smarter until some unknown inherent ceiling is reached? This process should be aided, in part, by educational systems that challenge children to learn, think, and create. Of course, intelligence, even genius, can appear from all quarters, educated or otherwise, because an enhanced mental potential can also have non-epigenetic origins (e.g., a favourable combination of parents' genes). Clearly, epigenetics supplements, not supplants, classical genetics. But epigenetics has important advantages over classical genetics in that epigenetics responds directly to specific needs and, compared to mutational events, epigenetic changes can occur more quickly. The obvious question to ask is, is artificial intelligence Darwinian or Lamarckian?

If education is one of the many mechanisms that catalyse an epigenetic release via the cranial feedback mechanism, then there exists an important conclusion that society should keep in mind. Comfort can be taken in knowing that teachers are moulding the intelligence not only of the current generation but of future ones as well. Since universal education has just begun, relative to our evolutionary timescale, its consequences will likely become more evident as time passes. In any event, neo-Lamarckian effects are yet another reason for establishing high-quality educational systems as a prime goal for all of humanity.

Lamarck's view of evolution, based on inheritance of acquired traits is opposed by neo-Darwinism, which instead is based on random DNA mutations. Thus, we are smarter than demanded by our evolutionary experience as hunter-gatherers. The difficulty lies in the inability of neo-Darwinism to satisfactorily answer the following question: How can a large energy-costly set of genes, each member of which has little apparent benefit when first created individually, all gather into a permanent existence within a short time period in each and every member of a small population (that was dispersed and geographically isolated over a huge planet) who had a low reproductive output, a low rate of beneficial mutations, and a low level of genetic contact? The article concludes with a speculative but far-reaching epigenetic theory of intelligence that does not require DNA mutation as the exclusive source of evolutionary change. Instead, cranial feedback relating brain chemistry, as affected by brain activity including education, with the genome. When it comes to the fast rate of evolution, and the dissemination of the intelligence trait worldwide, cranial feedback could make all the difference. Charles Darwin has been quoted:

> *The mystery of the beginning of all things is insoluble by us; and I for one must be content to remain an agnostic.*

> *If the misery of the poor be caused not by the laws of nature, but by our institutions, great is our sin.*

Emily Singer (2009) has reported two studies showing that the effects of a mother's early environment can be passed on to the next generation.

The effects of an animal's environment during adolescence can be passed down to future offspring. If applicable to humans, the research, done on rodents, suggests that the impact of both childhood education and early abuse could span generations. The findings provide support for a 200-year-old theory of evolution that has been largely dismissed: Lamarckian evolution, which states that acquired characteristics can be passed on to future generations.

In recent years, scientists have discovered that epigenetic changes – heritable changes that do not alter the sequence of DNA itself – play a major role in development, allowing genetically identical cells to develop different characteristics; epigenetic changes also play a role in cancer and other diseases. (The definition of epigenetics is somewhat variable, with some scientists limiting the term to refer to specific molecular mechanisms that alter gene expression.) Most epigenetic studies have been limited to a cellular context or have looked at the epigenetic effects of drugs or diet in utero. These two new studies are unique in that the environmental change that triggers the effect – enrichment or early abuse – occurs before pregnancy. "Give mothers chemicals, and it can affect offspring and the next generation," says Larry Feig, a neuroscientist at Tufts University School of Medicine, in Boston, who oversaw part of the research. "In this case, [the environmental change] happened way before the mice were even fertile."

In Feig's study, mice genetically engineered to have memory problems were raised in an enriched environment – given toys, exercise, and social interaction – for two weeks during adolescence. The animals' memory improved – an unsurprising finding, given that enrichment has been previously shown to boost brain function. The mice were then returned to normal conditions, where they grew up and had offspring. This next generation of mice also had better memory, despite having the genetic defect and never having been exposed to the enriched environment.

The researchers also looked at a molecular correlate of memory called **long-term potentiation**, or LTP, a mechanism that strengthens connections between neurons. Environmental enrichment fixed faulty LTP in mice with the genetic defect; the fixed LTP was then passed on to their offspring. The findings held true even when pups were raised by memory-deficient mice that had never had the benefits of toys and social interaction. "When you look at offspring, they still have the defect in the protein, but they also have normal LTP," says Feig.

"If the findings can be conveyed to humans, it means that girls' education is important not just to their generation but to the next one," says Moshe Szyf of McGill University, in Montreal, who was not involved in the research.

In a second study, researchers found that rats raised by stressed mothers that neglected and physically abused their offspring showed specific epigenetic modifications to their DNA. The abused mice grew up to be poor mothers, and appeared to pass down these changes to their offspring.

Previous research has shown that bad rat mothering can be passed down through this kind of DNA modification – but those changes are thought to be triggered specifically by maternal behaviour. In the new study, researchers also had healthy mothers raise the offspring of stressed mothers, and found that the problems were only partially fixed. That suggests that the changes "were not due to their neonatal experience," says David Sweatt a neuroscientist at the University of Alabama in Birmingham, who oversaw the study. "It was something that was already there when they were born."

The results of both studies are likely to be controversial, perhaps resurrecting a centuries-old debate. "It's very provocative," says Lisa Monteggia, a neuroscientist at the University of Texas Southwestern

Medical Centre, in Dallas. "It goes back to two schools of thought: Lamarck versus Darwin."

In contrast to natural selection, in which organisms that are born well-adapted to their environment survive and reproduce, passing down those successful traits, Lamarckian evolution suggests that animals can develop adaptive traits, such as better memory, during their lifetimes, and pass on those traits to their offspring. The latter theory was largely abandoned as Darwin's, and later Mendel's, theories took hold. But the concept of Lamarckian inheritance has made a comeback in recent years, as scientists learn more about epigenetics.

"I didn't set out to come up with findings that support neo-Lamarckian inheritance," says Sweatt. "But the research now makes it more plausible that these things may be real and may be based in molecular mechanisms."

Feig, on the other hand, argues that while the findings are "a Lamarckian kind of phenomenon it's still Darwinian, because the changes don't last forever." In Feig's study, the offspring of enriched mice lost their memory benefits after a few months.

Sweatt and others say that this type of inheritance may in fact be much more common than expected. Improving technologies are now providing a broader look at the epigenetic changes linked to environment and behaviour. Scientists are starting to use DNA microarrays, which over the past few years have become widely employed in genetic studies of disease, to look at one specific type of change, known as DNA methylation. "The changes we see are not limited to a small number of genes," says Szyf, who is using the technology to study epigenetics and cancer. "Whole circuitries are changed."

DNA sequencing, which is rapidly dropping in price, can also be used to study DNA methylation. But epigenetics studies require high-volume sequencing, which has been prohibitively expensive. "In contrast to the genome, every epigenome is different in different types of cells," says Sweatt. "A human epigenome project would be the equivalent of 250 human genomes, because there are at least 250 cell types in the body." Cheap sequencing may soon make that type of study possible, he says.

The actual mechanism underlying these patterns of inheritance is somewhat mystifying to scientists. Feig theorizes that environmental enrichment triggers a long-lasting hormonal change: when the animal becomes pregnant, the hormone would somehow modify the DNA of the foetus, ultimately causing it to have improved memory and LTP as an adolescent. However, he cautions, there is no direct evidence of this, and no specific evidence that the behaviours are transmitted through epigenetic mechanisms.

In 1972, paleontologists Niles Eldredge and Stephen Jay Gould published a landmark paper developing their theory that evolutionary biology utilises *punctuated equilibrium*. This proposes that once a species appears in the fossil record, the population will become stable, showing little evolutionary change for most of its geological history. This state of little or no morphological change is called *stasis*. When significant evolutionary change occurs, the theory proposes that it is generally restricted to rare and geologically rapid events of branching speciation called *cladogenesis* which is the process by which a species splits into two distinct species, rather than one species gradually transforming. Eldredge and Gould proposed that the degree of gradualism commonly attributed to Charles Darwin is virtually non-existent in the fossil record.

A perfect Lamarckian model was that elaborated by Cairns in 1988. Bacteria which were dependent on alteration of their RNA in order to utilise a new nutrient, reacted by producing more of this modified form of mRNA, which was preferentially retrotranscribed into the genome and transmitted to the progeny.

The neo-Darwinists suggested that evolution is constant and gradual, and thus that genetic changes that drive evolution should be too. However, more recent understanding of phenomena called adaptive mutation in microbes indicates that mutation rates can be elevated in response to stress, producing beneficial and other mutations.

Epigenetics is similarly considered Lamarckian, since epigenetic marks can be modified by the environment and these modifications (in rare cases) can be transmitted to the progeny.

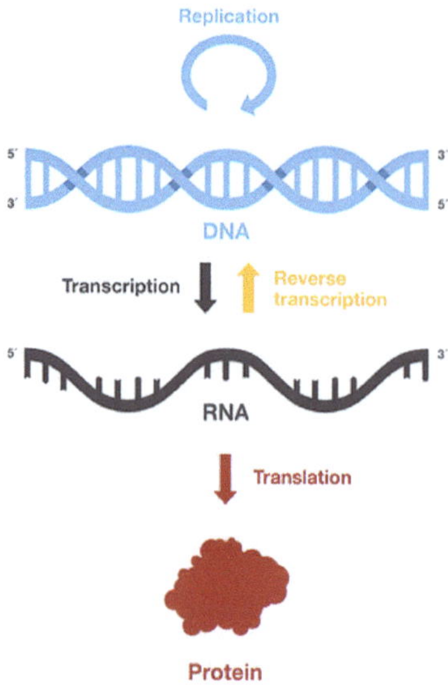

Give Chance a chance (1) that it may chance (2) upon the right chance (3).
1 = Creation, Big Bang, or whatever
2 = Darwinian evolution
3 = Lamarckism

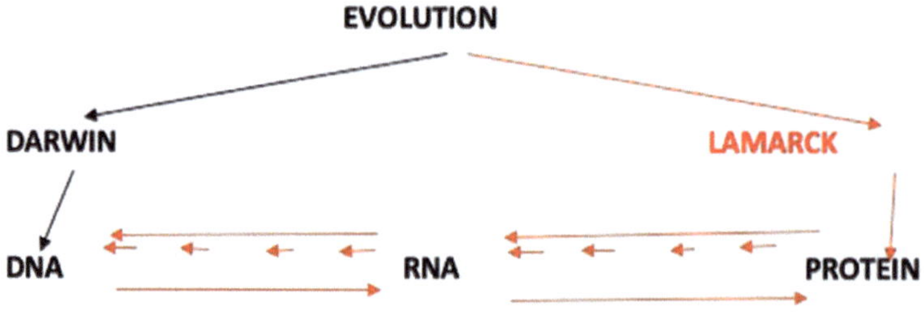

Evolution conjures images of physical transformation from common ancestor to present day forms. But there have been other invisible events, such as the acquisition of the use of tools and speech by Man. What has maybe gone unnoticed is the development of intelligence, since the brain may not show any obvious change.

Einstein's brain was removed at autopsy in 1955 and was found to be of unexceptional size. Its combination of a relatively wide and forward-projecting right frontal lobe with a relatively wide and posteriorly protruding left occipital lobe is the most prevalent pattern seen in right-handed adult males. The morphology in some parts of Einstein's cerebral cortex was highly unusual compared with some other human brains for which sulcal patterns have been thoroughly described. However, before jumping to conclusions about these features, though admittedly different, they were only compared with a total of 85 other human brains, which makes it difficult to draw any firm conclusions about their significance.

An international team of researchers at Harvard University with partial support from the National Institute of Biomedical Imaging and Bioengineering (NIBIB) developed a new MRI technique that could capture an image of a brain thinking by measuring changes in tissue stiffness. The results showed that brain function could be tracked on a timescale of 100 milliseconds – 60 times faster than previous methods. The technique might shed new light on altered neuronal activity in brain diseases.

The human brain responds almost immediately to stimuli, but non-invasive imaging techniques haven't been able to keep pace with the brain. Currently, several non-invasive brain imaging methods measure brain function, but they all have limitations. Most commonly, clinicians and researchers use functional magnetic resonance imaging (fMRI) to measure brain activity via fluctuations in blood oxygen levels. However, a lot of vital brain activity information is lost using fMRI because blood oxygen levels take about six seconds to respond to a stimulus.

Since the mid-1990s, researchers have been able to generate maps of tissue stiffness using an MRI scanner, with a non-invasive technique called magnetic resonance elastography (MRE). Tissue stiffness can't be measured directly, so instead researchers use MRE to measure the speed at which mechanical vibrations travel through tissue. Vibrations move faster through stiffer tissues, while vibrations travel through softer tissue more slowly; therefore, tissue stiffness can be determined. MRE is most commonly used to detect the hardening of liver tissue but has more recently been applied to other tissues like the brain.

"This study has the potential to revolutionize the way scientists study brain diseases," says Krishna Kandarpa, M.D., Ph.D., director of research science and strategic directions at the National Institute of Biomedical

Imaging and Bioengineering, at Bethesda, Maryland, USA. "Developing a new MRI technique relies heavily on physics and engineering principles, which are areas in which NIBIB investigators excel. The results would have been hard to achieve without the collaboration of this team of experts."

This form of study and future ones would be of immense value particularly when correlated with IQ and other indications of exceptional intelligence.

Clock speed is the rate at which a computer's processor can complete a processing cycle. It is typically measured in megahertz or gigahertz. One megahertz is equal to 1 million cycles per second, while one gigahertz equals 1 billion cycles per second.

The UNIVAC mainframe computer in 1951 consisted of 5,200 vacuum tubes (almost all in the processor), weighed 29,000 pounds, consumed 125 KW, and ran at a whopping 2.25 MHz clock speed, 46 units of which were sold to businesses and government. It was capable of doing 455 multiplications per second.

The UNIVAC was the first-ever commercial mainframe computer in 1951.
Clock speed 2.25 Mhz

Dell Inspiron laptop, 2020. Clock speed up to 3.4 GHz

It can be seen that the laptop above is 1,500 times faster than the 1951 Univac! Obviously, size isn't everything! Similarly, why should human brains, the object of evolving intelligence, necessarily get larger?

Samuel Morton was a 19th-century anthropologist whose work on skulls of different races has been a source of debate for over a century. Morton measured the volume of black and white human skulls, concluded that white skulls had a greater volume, and on this evidence claimed that this undoubtedly accounted for their superior intelligence. However, the sheer size of a human brain need not equate to intelligence. In fact some people with microcephaly (abnormally small skull) have normal intelligence. And let us not forget that the combined effect of scientific knowledge, education, diet, and improvements in medicine, may have had an effect on the development and improvement in the acquisition of knowledge. Intelligence may also enjoy improvement as part of the evolutionary process. Currently great strides are being made in the field of artificial intelligence under the guidance and control of humans.

Lamarckism supposes that species acquire characteristics to deal with challenges experienced during their lifetimes, and that such accumulations are then passed to their offspring. In modern terms, this transmission from parent to offspring could be considered a method of epigenetic inheritance. Scientists are now questioning the framework of the modern synthesis, as epigenetics to some extent is Lamarckist rather than Darwinian. While some evolutionary biologists have dismissed epigenetics' impact on evolution entirely, others are exploring a fusion of epigenetic and traditional genetic inheritance.

There is now a body of evidence that immunity is in part genetically regulated and that evolution has played a part in its increased sophistication. This applies to both innate and acquired immune responsiveness, involving immune response genes both in mice and humans and though initially extensively researched in mice (the H-2 system) have subsequently received considerable effort in elucidating the nature of the genetics of human responsiveness. The human equivalent of the mouse H-2 system, is the MHC, which stands for major histocompatibility complex, the major relevance of which is in predicting the successful grafting of organs from other human beings. Kidney transplants are now commonplace, also, heart, heart-lung, and liver. Graft rejection is mainly determined by non-identical tissue type. Success depends in large part on immune suppression by drugs.

Some diseases are associated with certain genes, for example ankylosing spondylitis and HLA-B27, and rheumatoid arthritis, HLA-DR4. Immune deficiency can be genetically inherited.

We have recently been in the midst of a global pandemic due to the corona virus, and it is becoming increasingly clear that there are some tantalizing facts emerging, in regard to susceptibility, and mortality. At one end is the observation that some people test positive by detection of antigen in their throat or nose, yet have no symptoms whatever, others a relatively mild illness, from which a rapid and complete recovery occurs within weeks. At the opposite end of the spectrum, is a rapid increase in respiratory symptoms, with a severe pneumonia, and eventually, despite all attempts, including the use of respirators, death. In addition to these facts, is the increased risk to the elderly, and non-white races. Whether the latter is explicable on the basis of socio-economic differences is currently receiving much attention by the media. Intriguingly, it appears that an exaggerated immune response as contrasted with an impaired one may be responsible for the most serious cases.

The most sophisticated immune systems are those of the vertebrates. Recognizable lymphocytes and immunoglobulins (Ig; also called antibodies) appear only in these organisms. The most primitive living vertebrates – the jawless fishes (hagfish and lampreys) – do not have lymphoid tissues corresponding to a spleen or a thymus, and their immune responses, although demonstrable, are very weak and sluggish. Further up the evolutionary tree, at the level of the cartilaginous fishes (sharks and rays) and the bony fishes, a thymus and a spleen are present, as are immunoglobulins, although only those immunoglobulins of the IgM class are detectable. Fish lack specialised lymph nodes, but they do have clusters of lymphocytes in the gut that may serve an analogous purpose.

Not surprisingly, molecules that have a similar function in different species (e.g., immunoglobulins or MHC components) show an even closer resemblance. By analysing the number of differences in the amino acids and their position in the polypeptide chains that make up IgM and IgG molecules in, for example, humans, mice, and rabbits and by making reasonable assumptions about mutation rates, scientists can estimate roughly how many generations would have had to elapse – and hence how much time – for the present immunoglobulins of these species to have evolved from a common ancestral IgM-like molecule. Such calculations suggest that divergence from the ancestral immunoglobulin took place some 200 million years ago. That was about the same time amphibians are thought to have diverged from the main vertebrate line. So one may conclude that a functional immune system arose even earlier and has continued to provide a defence against foreign agents ever since.

On the negative side of the ledger are the policies derived from the belief in the inevitability of a Malthusian collapse.

> *The power of population is so superior to the power of the earth to produce subsistence for man, that premature death must in some shape or other visit the human race.*

Thus Malthus gloomily predicted. His scenario influenced policy makers to embrace social Darwinism and eugenics, resulting in draconian measures to restrict particular populations' family size, including forced sterilizations.

In his book *The Evolution of Everything* (Harper, 2015), evolutionary biologist and journalist Matt Ridley summed up the policy succinctly: "Better to be cruel to be kind." The belief that "those in power knew best what was good for the vulnerable and weak" led directly to legal actions based on questionable Malthusian science. For example, the English Poor Law implemented by Queen Elizabeth I in 1601 to provide food to the poor was severely curtailed by the Poor Law Amendment Act of 1834, based on Malthusian reasoning that helping the poor only encourages them to have more children and thereby exacerbate poverty. The British government had a similar Malthusian attitude during the Irish Potato Famine of the 1840s, Ridley notes, reasoning that famine, in the words of Assistant Secretary to the Treasury Charles Trevelyan, was an "effective mechanism for reducing surplus population." A few decades later Francis Galton advocated marriage between the fittest individuals ("What nature does blindly, slowly, and ruthlessly Man may do providently, quickly and kindly"), followed by a number of prominent socialists such as Sidney and Beatrice Webb, George Bernard Shaw, Havelock Ellis and H. G. Wells, who openly championed eugenics as a tool of social engineering.

We think of eugenics and forced sterilization as a right-wing Nazi program implemented in 1930s Germany. Yet as Princeton University economist Thomas Leonard has documented in his book *Illiberal Reformers* (Princeton University Press, 2016) and former *New York Times* editor Adam Cohen reminded us in his book *Imbeciles* (Penguin, 2016), eugenics fever swept America in the early 20th century, culminating in the 1927 Supreme Court case Buck v. Bell, in which the justices legalized sterilization of "undesirable" citizens. The court included prominent progressives Louis Brandeis and Oliver Wendell Holmes, Jr., the latter of whom famously ruled, "Three generations of imbeciles are enough." The result: sterilization of some 70,000 Americans!

Science writer Ronald Bailey tracks neo-Malthusians in his book *The End of Doom* (St Martin's Press, 2015), starting with Paul Ehrlich's 1968 best seller *The Population Bomb*, which proclaimed that "the battle to feed all of humanity is over." Many doomsayers followed. Worldwatch Institute founder Lester Brown, for example, declared in 1995, "Humanity's greatest challenge may soon be just making it to the next harvest." In a 2009 Scientific American article he affirmed his rhetorical question, "Could food shortages bring down civilization?" In a 2013 conference at the University of Vermont, Ehrlich assessed our chances of avoiding civilizational collapse at only 10 percent.

The problem with Malthusians, Bailey writes, is that they "cannot let go of the simple but clearly wrong idea that human beings are no different than a herd of deer when it comes to reproduction." Humans are thinking animals. We find solutions. The result is the opposite of what Malthus predicted: the wealthiest nations with the greatest food security have the lowest fertility rates, whereas the most food-insecure countries have the highest fertility rates.

The solution to overpopulation is not to force people to have fewer children. China's one-child policy showed the futility of that experiment. It is to raise the poorest nations out of poverty through democratic governance, free trade, access to birth control, and the education and economic empowerment of women.

It is tempting to postulate that whereas food shortage and its effect on an ever-increasing population, evoked the well accepted Malthusian response, we are nowadays seeing a threat to the human race posed by the increased demands of capitalism, which has contributed to global warming. Thus for yesterday's (1845) potato famine substitute today's monetary famine and the consequence of the coronavirus pandemic which by wrecking the planet's monetary systems, will set the world economy at a different and hopefully more sustainable level. *Per virus planetae salvare.* (Through the virus survival of the planet).

Richard Dawkins, who previously alluded to (1976) the *selfish gene* concept, has coined the term *meme* as an idea, behaviour, or style that spreads by means of imitation from person to person within a culture and often carries symbolic meaning representing a particular phenomenon or theme. A meme acts as a unit for carrying cultural ideas, symbols, or practices, that can be transmitted from one mind to another through writing, speech, gestures, rituals, or other imitable phenomena with a mimicked theme. Supporters of the concept regard memes as cultural analogues to genes in that they self-replicate, mutate, and respond to selective pressures.

Proponents theorize that memes are a viral phenomenon that may evolve by natural selection in a manner analogous to that of biological evolution. Memes do this through the processes of variation, mutation, competition, and inheritance, each of which influences a meme's reproductive success. Memes spread through the behaviour that they generate in their hosts. Memes that propagate less prolifically may become extinct, while others may survive, spread, and (for better or for worse) mutate. Memes that replicate most effectively enjoy more success, and some may replicate effectively even when they prove to be detrimental to the welfare of their hosts. (Thank goodness the Nazi meme did not succeed).

Epigenetic modifications can turn genes on and off by affecting the chemical structure of the DNA or its associated proteins, rather than the DNA sequence. Unlike genetic modifications, epigenetic changes are transient and reversible.

Mortality rates in influenza appear to have been shaped by evolution. During the 1918 pandemic, mortality rates were lower in children compared with adults. This mortality difference occurs in a wide variety of infectious diseases. It has been replicated in mice and might be due to *greater tolerance of infection, not greater resistance*. Treating the host response by damping it down might be a practical way to reduce global mortality during the current pandemic. It might also help reduce mortality due to seasonal influenza and other forms of acute critical illness.

Influenza virologists have expanded our understanding of the molecular biology and epidemiology of influenza viruses. Laboratory and clinical investigators have deepened our understanding of the host response to critical illness. Evolutionary biologists have suggested that evolution provides insights that could help public health. All of these developments should shape the way we respond to the next influenza pandemic.

Charles Darwin wrote "... observation must be for or against some view if it is to be of any service". His view – the hypothesis that evolution is guided by natural selection – was supported by extensive observations made over several decades, and its explanatory power (i.e., 'service') has withstood challenge for almost 160 years. The idea (hypothesis) of treating the host response to an influenza pandemic was introduced in 2004. Its potential explanatory power has also been supported by experimental and clinical observations, although treating the host response has received little attention from scientists and health officials. Yet in a practical sense this approach to treatment could be of great service (in Darwin's word) if it could reduce global mortality during the next pandemic, *which as it happens is now upon us.*

Kobzik and colleagues have shown age-specific differences in influenza mortality in mice before and after puberty that are not affected by previous infection (antigenic imprinting) with influenza viruses. ***Their findings suggest that the mortality impact of pandemic and seasonal influenza and other forms of acute critical illness might be reduced by treating the host response. By reducing the damage caused by influenza in adults to the more tolerant response seen in children, treatment could in effect "roll back" evolution.*** *This is becoming more apparent with the realization that deaths due to Cov-19 might be the consequence of so-called cytokine storm.*

The biological basis for treating the host response reflects our evolutionary heritage. This idea might not be revolutionary, but its practical implications for health, equity and security during the next pandemic could be immense. Consequently, investigators must undertake laboratory and clinical research to convincingly show whether it will be effective. If it is effective and is put into practice, it would represent a striking application of the idea of evolutionary public health.

Why may ancient pandemics hold the key to our future survival?

Our DNA bears the scars of an arms race that humanity has been fighting for millions of years.

Twenty-five thousand years ago, the first elements of modern humanity were beginning to emerge in Europe. One of the earliest known settlements, near the Dyje River in what is now the Czech Republic, was founded; a place where people lit fires, spun thread, wore clothes and caught fish on hooks. Ceramics had just been developed, and people began to form clay or to carve other materials into likenesses of men, women and deities. Humans had reached an important new stage on the journey that had begun 2 million years earlier with the sharpening of the first flint knife.

Then, in large numbers, they began to fall ill.

How can we say that a virus caused a pandemic 20,000 years before the beginning of recorded history? David Enard, an evolutionary biologist who works at the University of Arizona, says the scars of ancient epidemics, and the history of our long and ceaseless war with viruses, are written in our DNA. Finding out what caused them may help us to predict the next pandemic.

Enard's work involves searching human genomes for evidence of ancient epidemics. It has led him to believe the current outbreak is one in a series of hundreds or even thousands of epidemics and pandemics that have afflicted humans through every moment of our evolution. In 2016, while he was at Stanford University, Enard showed that around 30% of all the adaptive mutations in our genome – a third of human evolution – has been in response to viruses.

Even within the span – vanishingly brief, in evolutionary terms – of recorded history there are examples such as the Black Death, caused by the bacterium *Yersinia pestis*, the devastating influenza pandemics of 1889 and 1918, and the plagues that ravaged the Greeks, Romans, Aztecs and many others. "But if you start thinking in terms of hundreds of thousands of years of evolution... it leaves a lot of time for hundreds or even thousands of pandemics to have occurred."

Enard finds the evidence of these ancient plagues by looking at the effect they had on our evolution. Early people would have had some knowledge of medicine, but this would have done nothing to prevent or cure a new infectious disease. The only defence they had was that which they carried in their genes.

"Mutations are usually harmful," Enard explains, ***"but when a pandemic happens, it could be that some individuals... have mutations that actually make them resistant against the infection."***

There are several ways that genes could confer an advantage. One kind of mutation could be a change in the immune system, which would "make the immune cells of these people more able to control the virus early on, so that the virus does not multiply so much". ***The RSAD2 gene*** that most mammals possess, for example, causes cells to produce a substance that prevents viruses from copying their genomes and replicating. It is thanks to this gene, among many others, that we do not succumb to the trillions of viral particles that cover our skin, infest our noses and digestive systems, and wash through our blood. This is due to the genetically encoded *virus inhibitory protein*, endoplasmic reticulum-associated, interferon-inducible ***(Viperin)***, also known as RSAD2. Viperin is a multifunctional protein in viral processes that is an interferon (IFN) stimulated gene. It has been reported that viperin could be induced by either IFN-dependent or IFN-independent pathways and certain viruses may use viperin to increase their infectivity.

Viperin is normally localized to the endoplasmic reticulum (ER) via its N-terminal domain, and also localized to lipid droplets, which are derived from the ER. However, it is also found in mitochondria in the HCMV infected fibroblasts.

The other way in which our genes could adapt to a virus would be by controlling the immune response. In the case of Covid-19, for example, the virus provokes a reaction – a ***"cytokine storm"*** – that is so strong that it seriously damages or even kills the body it was supposed to protect. But in some cases, says Enard, mutations have allowed people to "build a more controlled immune response" to viruses, allowing them to endure their own defences for long enough to recover.

It is believed that cytokine storms were responsible for the disproportionate number of healthy young adult deaths during the 1918 influenza pandemic, which killed 17 to 50 million people. In this case, a healthy immune system may have been a liability rather than an asset. Preliminary research results from Taiwan also indicated this as the probable reason for many deaths during the SARS epidemic in 2003. Human deaths from the bird flu H5N1 usually involve cytokine storms as well.

In 2006, a study at Northwick Park Hospital in England resulted in all 6 of the volunteers given the drug theralizumab becoming critically ill, with multiple organ failure, high fever, and a systemic inflammatory response. Parexel, a company conducting trials for pharmaceutical companies, in one of its documents, wrote about the trial and said theralizumab could cause a cytokine storm – the dangerous reaction the men experienced.

How does he know which virus was seen off by a particular mutation? When a viral particle, or virion, comes into contact with a human cell, its proteins interact physically with the proteins of the cell. It is, says Enard, as if the human cell is a factory, and the virus breaks in, takes specific pieces of equipment – proteins – and reprogrammes them to make copies of itself.

"We happen to know, thanks to the hard work done by virologists, which specific human proteins interact with which specific viruses," he explains. "There are around 20,000 human proteins in human genomes in total. And we know from work that has been published very recently that Covid-19, for example, seems to interact with about 300 of those 20,000 proteins."

If Enard and his fellow researchers see a mutation that is relevant to these 300 specific proteins, it suggests that the mutation was relevant to a coronavirus. And if this mutation can be seen to have spread across many generations of people, over thousands of years, it suggests that this gene may have been carried by people who survived a coronavirus pandemic in the prehistoric world.

To figure out when such a pandemic happened, Enard uses population genetics, a science that he compares to the building of family trees, but on a vast scale. By analysing the genomes of large numbers of people from around the world, and the differences between them, population genetics tools can build genealogies that "reconstruct the genealogy of those people on the very deep timescale" – drawing trees of relations that extend tens of thousands of years into the past.

When Enard combines the data from population genetics with the mutations that have allowed people to defeat viruses, there emerge "sweeps" – peaks in the data that signify where a mutation became important to survival. Looking at RNA viruses – which include coronaviruses, *ebola*, *zika* and others that we know have caused new infectious diseases – he has seen signs of very significant sweeps between 300 and 900 generations back. Somewhere around 10,000 and 25,000 years ago, RNA viruses "imposed a strong pressure on the ancestors of a specific human population at the time that the advantageous mutations emerged".

Evolutionary biologists use phrases such as "selective pressure" or "evolutionary pressure" to describe factors that make it more difficult to pass on one's genes. This does not always mean death – Enard explains that evolution can be counterintuitive, in that even very small advantages can "trigger strong and rapid natural selection" – but in the case of a disease, the strongest selective pressure is exerted when the pathogen kills everyone who isn't immune to it. This is the sombre truth of evolution. "When we talk about adaptation, it sounds like some something positive, right? But it happens because there's something very bad happening."

Chapter 13

Evolution and the Role of Stress

Bust of Hans Selye at Selye János University, Slovakia (Wikipedia)

Hans Selye was born in 1907 in Vienna, and died in 1982, in Montreal. His name is widely recognised for his pioneering work on stress. He became a Doctor of Medicine and Chemistry in Prague in 1929 and went on to do pioneering work in stress and endocrinology at Johns Hopkins University, McGill University, and the Université de Montréal. He was nominated for the Nobel prize in physiology or medicine for the first time in 1949. Although he received a total of 17 nominations in his career, he never won the prize.

A pioneering Hungarian-Canadian endocrinologist, he conducted important scientific work on the hypothetical non-specific response of an organism to whatever led to stress. This he termed the *stressor*. Although he did not recognise all of the many aspects of steroids, Selye was aware of their role in the stress response.

His interest in stress began when he was in medical school; he had observed that patients with various chronic illnesses like tuberculosis and cancer appeared to display a common set of symptoms that he attributed to what is now commonly called stress. After completing his medical degree and a doctorate degree in organic chemistry at the German University of Prague, he received a Rockefeller Foundation fellowship to study at Johns Hopkins in Baltimore and later moved to the Department of Biochemistry at McGill University in Montreal where he studied under the sponsorship of James Bertram Collip. While working with laboratory animals, Selye observed a phenomenon that he thought resembled what he had previously seen in chronic patients. Rats exposed to cold, drugs, or surgical injury exhibited a common pattern of responses; this he called the *general adaptation syndrome* or GAS. It had 3 stages, involving an initial alarm phase followed by a stage of resistance or adaptation and, finally, a stage of exhaustion and death. Working with doctoral student Thomas McKeown, Selye published a report that used the word "stress" to describe these responses to adverse events.

His last inspiration for general adaptation syndrome came from an endocrinological experiment in which he injected mice with extracts of various organs. He at first believed he had discovered a new hormone, but was proved wrong when every irritating substance he injected produced the same features (swelling of the adrenal cortex, atrophy of the thymus, gastric and duodenal ulcers). The actual hormone was cortisone, produced in the adrenal cortex. This, paired with his observation that people with different diseases exhibit similar symptoms, led to his description of the effects of "noxious agents" as he at first called it. He later coined the term *stress,* which has been accepted in most other languages.

Survival stress differs from other physical responses in that stress is stressful whether one receives good or bad news, whether the impulse is positive or negative. He called negative stress *distress* and positive stress *eustress*. The system whereby the body copes with stress, the hypothalamic-pituitary-adrenal axis (HPA axis) system, was also first described by Selye. He also pointed to an "alarm state", a "resistance state", and an "exhaustion state", largely referring to glandular states. Later he developed the idea of two "reservoirs" of stress resistance, or alternatively stress energy. The adrenal glands as their names imply, are situated over the kidneys. They consist mainly of the *cortex*, which surrounds the inner area, called the *medulla.* The cortex produces several types of steroids, which are involved in stress reactions such as infections, whereas the medulla produces adrenalin and is fast acting for fight or flight.

Evolution in the Darwinian sense, evokes images of species divaricating, and fortuitously gaining a toe-hold permitting greater longevity. However, survival to later life brings with it the appearance of chronic diseases. Evolution requires reproduction, and mostly nurturing of young, but when that time is exceeded, there is less evolutionary pressure, hence the emergence of said diseases, e.g. hypertension, and diabetes. Emotional stress has a dual side, one being biologically negative and pathogenic, the other by adaptation leading to self-preservation and eventually, evolutionary change within species. The Dynamic theory of emotions describes the appearance of a continuous negative emotional state in conflicting behavioural situations when humans and animals cannot satisfy their leading biological or social needs. **Youmatov** working in Moscow in 1997 produced experimental evidence of this by the use of rats, which were confined in tight cages, and displaying emotional stress. By contrast other rats, fleeing from electric shocks, but then into similar confined cages did not show signs of stress.

Dona Lee Wong working from Harvard University in 2006, wrote that stress contributes to the pathophysiology of many diseases, including psychiatric disorders, immune dysfunction, nicotine addiction and cardiovascular illness. The major stress hormones are epinephrine (adrenaline) and the glucocorticoids, cortisol and corticosterone.

The release of adrenaline from the adrenal medulla and glucocorticoids from the adrenal cortex initiate the biological responses permitting the organism to cope with adverse psychological, physiological and environmental stressors. Following its massive release during stress, adrenaline must be restored to replenish cellular pools and sustain release to maintain the heightened awareness and sequelae of responses to re-establish homeostasis and ensure survival.

Adrenaline is regulated in part through its biosynthesis catalysed by the final enzyme in the catecholamine pathway, phenylethanolamine

N-methyltransferase (PNMT). PNMT expression, in turn, is controlled through hormonal and neural stimuli, which exert their effects on gene transcription through protein stability.

The pioneering work of **Julius Axelrod** forged the path to our present understanding of how the stress hormone and neurotransmitter epinephrine, is regulated, in particular via its biosynthesis by PNMT.

Youmatov (2020) has drawn attention to the positive role of emotional stress:

> *biological evolution is a natural process of development of wild life, accompanied by a change in the genetic composition of populations, the formation of adaptations, speciation and extinction of species, the transformation of ecosystems and the biosphere as a whole.*

He concludes that emotional stress is one of the factors in the modern evolution of species. This is contained within the *Dynamic Theory of Emotions* which characterises the development of a continuous negative emotional state in conflicting behavioural situations when the subject is unable to satisfy its strong dominant need for some time. The emotional stress is represented as having a dual nature i.e. a biologically negative pathogenic effect on health, but a positive impact on the adaptation of individuals, self-preservation of life, and evolutionary change in species.

With increasing medical knowledge and treatment of infectious disease, plus improvements in sanitation, the challenge, or first act, posed by infectious diseases has, to a degree, been modified, though admittedly some glaring contradictions remain, e.g. malaria, and viral disease, currently coronavirus Covid-19. The second act has seen the emergence of chronic degenerative diseases, e.g. hypertension, ischaemic heart disease, and diabetes. In the main, though not exclusively, these conditions are less influential on reproduction in view of their relatively late onset. Nevertheless, despite this, there are genetic variations detectable influencing their effect. These observations have emerged from the study of 'race' in various populations. However, the confounder here has been shown to be the role of poverty and poor diet and lifestyle. It is interesting that artificial selection in rats and mice can result in analogous degenerative disorders e.g. hypertension, obesity and various cancers, which has been interpreted as implying genetic sharing with the aforementioned species, and by conclusion having an evolutionary origin prior to the divergence of humans and rodents.

In the future, the mechanisms of stress-induced evolutionary effects will be clarified by research into genes ('genomics').

Chapter 14

Future Evolution

The following text is partly taken from Google search.

Where is evolution taking us? Insofar as the human race is concerned, the question 'Are we still evolving, or have we reached the pinnacle?' must be asked. The fossil record tells us that the oldest members of our species lived 195,000 years ago in what is now Ethiopia. From there *Homo sapiens* spread out across the globe. Ten thousand years ago humans had colonized all the continents save Antarctica. Adaptations to life in these regions might have been expected to lead to the development of different species on account of geographical separation, but it did not. This would imply that there was contact across all these regions. So, although it didn't result in new species being formed, it did result in the development of racial variation, as currently perceived.

Darwin's *On the Origin of Species*, noted the differences in the beak size and shape according to which islands these birds were found, and explained them on separation by distance so that mutations would be encouraged by lack of competition from the original forms. However, when travel between distant localities becomes an everyday fact, e.g. air travel, the likelihood of mutations gaining superiority becomes unlikely.

Steve Jones, a genetics professor at University College London, put forward a similar scenario during a lecture at the University of Cambridge marking the bicentenary of Darwin's birth and the 150th anniversary of

On the Origin of Species. "The human population will become more alike as races merge," he said, but:

> Darwin's machine has lost its power. That's because natural selection – Darwin's *survival of the fittest* concept – is being sidelined in humans. The fittest will no longer spearhead evolutionary change, because, thanks to medical advances, the weakest also live on and pass down their genes.

When *On the Origin of Species* was published in 1859, only about half of British children survived to 21. Today that number has swelled to 99 percent. In developed countries, "the fact that (nearly) everybody stays alive, at least until they're sexually mature, means 'survival of the fittest' has got nothing to work with. That part of the Darwinian fuel has gone."

However other scientists see plenty of evidence that human evolution is far from over. A study found that shorter, slightly plumper women tend to have more children than their peers. These physical traits are passed on to their offspring, suggesting natural selection in humans is alive and well.

What makes a woman chose a mate? Obviously male looks and physique play a part, but in today's world, materialism also plays a large part. Thus ostentatious wealth is a major 'pulling' point, though not necessarily to happiness. Wealth however may accrue from the application of intelligence, i.e. business savvy, which appeals to many women. Consequently, this intelligence may be passed on to the next generation.

Artificial selection using genetic technologies will likely accentuate these changes in the future, with parents choosing which sperm and egg to combine to produce a baby based on genetic information about which genes contribute to which desired physical and mental traits. As ever, market forces may determine which couples manage to produce offspring with the desired characteristics.

The human population will become more alike as races merge, because natural selection is being sidelined. As mentioned above the fittest will no longer spearhead evolutionary change, because, thanks to medical advances, the weakest also live on and pass down their genes. Thus the selection process which would have previously killed off the most physically (and in many cases, genetically) disadvantaged has led to far more surviving and reproducing.

It is likely that the rich and powerful will take advantage of the artificial-selection technology as a consequence of which there would be a split between a kind of upper-class, dominant population and a lower-class, genetically oppressed population.

With the advent of genetic manipulation it is theoretically possible that humans may, by transmission of their minds into computers incorporating artificial intelligence, achieve electronic immortality. A philosophy known as *transhumanism* sees humans taking charge of their evolution and transcending their biological limitations via technology.

In essence, the old-fashioned concept of evolution of the species involving natural selection may be beside the point: The future may belong to *unnatural selection*.

Nick Bostrom, director of the Future of Humanity Institute at the University of Oxford, has said Darwinian evolution:

> *is happening on a very slow time scale now relative to other things that are leading to changes in the human condition.*

Cloning, genetic enhancement, robotics, artificial intelligence, and nano-technology, for starters, which act much faster.

Transhumanism raises a spectacular array of possibilities, from supersoldiers and new breeds of athletes to immortal beings who, having had their brains scanned atom by atom, transfer their minds to computers.

In addition to living forever, "uploaded" beings would be able to travel at the speed of light as an information pattern, download themselves into robots for the occasional stroll through the real world, think faster when running on advanced operating systems, and cut their food budget down to zero.

If that were to happen, a new type of evolution would emerge. Thus:

> *Evolutionary selection could occur in a population of uploads or artificial intelligence just as much as it could in a population of biological organisms, in fact, it might operate much faster there, because artificial intellects could reproduce much faster.*

Whereas the current human generational cycle takes some 20 years, a digitalized individual could replicate themselves in seconds or minutes, and of course copying yourself isn't without complications, Bostrom acknowledges.

"Which one of them is you?" he writes. "Who owns your property? Who is married to your spouse?" (This thought is quite terrifying!).

The fact that speciation is known to have occurred as a consequence of separation in distance and time, in the far distant future, habitable planets beyond our solar system may become colonized by Earth migrants, who could provide the necessary isolation for new human species to evolve.

"If we had spacefaring people who went on one-way voyages to distant stars, that might be enough to trigger speciation," said John Hawks, anthropologist at the University of Wisconsin-Madison. However he continued:

> *If you think about it, a small group of people went on a one-way voyage to the Americas 14,000 years ago, and then when new people, Europeans showed up 500 years ago, they were still the same species.*

Same species, yes, but evolution can be ongoing in other ways only revealed by genetic analysis. For instance analysis of us Britons, has indicated that we are nowadays taller and fairer since Roman times, and intriguingly, that a gene associated with a propensity for cigarette smoking has dwindled over the course of just one generation. If the reduction is ascribed to less smoking, then might this not be regarded as smacking of Lamarckism as a consequence of withholding exposure to tobacco and consequently ceasing to favour selection of this gene? With access to giant genomic data sets, shifts in differing (allele) gene frequencies can be recorded over short timescales. The human genome quickly responds to new conditions in subtle but meaningful ways. Researchers at Harvard University found 300 regions on the human genome showing recent changes that improved the chances of surviving Lassa fever, and partial resistance to malaria in some Africans.

Evolutionary biologists have long concentrated on the role of new mutations in generating new traits. But once a new mutation has arisen, it must spread through a population. Every person carries two copies of each gene, but the copies can vary slightly within and between individuals. Mutations in one copy might increase height; those in another copy, or allele, might decrease it. If changing conditions favour, say, tallness, then tall people will have more offspring, and more copies of variants that code for tallness will circulate in the population. With the help of giant genomic data sets, scientists can now track these evolutionary shifts in allele frequencies over short timescales. Jonathan Pritchard and Yair Field at Stanford University, California did so by counting unique single-base changes, which are found in every genome. (To recap, the bases in DNA are either adenine, thymine, cytosine, or guanine, abbreviated A-T-C-G, and in RNA, thymine is replaced by uracil). Such rare individual changes, or *singletons*, are likely recent, because they haven't had time to spread through the population. Because alleles carry neighbouring DNA with them as they circulate, the number of singletons on nearby DNA can be used as a rough molecular clock, indicating how quickly that allele has changed in frequency. The more singletons, the longer the time period, and by contrast, fewer singletons mean less time, i.e. the evolutionary effect is more rapid. Pritchard's team analysed 3,000 genomes collected as part of the UK10K sequencing project in the United Kingdom. The more intense the selection on an allele, the faster it spreads, and the less time there is for singletons to accumulate near it. This approach can reveal selection over the past 100 generations, or about 2,000 years. Stanford University workers found relatively few singletons near alleles that confer lactose tolerance – a trait that enables adults to digest milk – and that code for particular immune system receptors. Among the British, these alleles have evidently been highly selected and have spread rapidly. The team also found fewer singletons near alleles for blond hair and blue eyes, indicating that these traits, too, have rapidly spread over the past 2,000 years. One evolutionary promoter may have been Britain's gloomy skies. Genes for fair hair also cause lighter skin colour, which allows the body to make more vitamin D in conditions of scarce sunlight. Or sexual selection could have been at work, driven by a preference for blond mates. Other researchers praise the new technique. "This approach seems to allow much more subtle and much more common signals of selection to be detected," says evolutionary geneticist Svante Pääbo of the Max Planck Institute for Evolutionary Anthropology in Leipzig. In a sign of the method's power, Pritchard's team also detected selection in traits controlled not by a single gene, but by tiny changes in hundreds of genes. Among them are height, head circumference in infants, and hip size in females – crucial for giving birth to those infants. By looking at the density of singletons flanking more than 4 million DNA differences, Pritchard's team discovered that selection for all three traits occurred across the genome in recent millennia. In a 2007 study by Harpending of the University of Utah, and Hawks of the University of Wisconsin-Madison, genetic markers of 270 people gathered from Han Chinese, Japanese, Yoruba and northern Europeans found that at least 7 percent of their genes had undergone evolutionary changes during the past 5,000 years. This was apparently due to changing environment, natural or man-made. In China and Africa, few people can digest milk once they reach adulthood, in contrast to Swedish and Danish populations. This is manifested as *lactose intolerance*, due to lack of the enzyme *lactase*. The gene that produces lactase is located on chromosome 2 in humans. Multiple chains are put together to form lactase, which is made up of four of the same subunits. Each subunit has 1,023 amino acid residues for a total of 4,092 amino acid residues. The environmental stimulus was presumably dairy farming.

Joseph Pickrell, an evolutionary geneticist at the New York Genome Center has used a different strategy to put selection under an even keener microscope, by detecting signs of evolution on the scale of a human lifetime. He and Przeworski studied the genomes of 60,000 people of European ancestry who had been genotyped by *Kaiser Permanente* (Kaiser Permanente is an American integrated managed care consortium in northern California), and 150,000 people from a massive UK sequencing effort called the *UK Biobank*. They wanted to know whether genetic variants change frequency across individuals of different ages, and revealed selection at work within a generation or two. The biobank included relatively few old people, but it did have information about participants' parents, so the team also looked for connections between parental death and allele frequencies in their children.

In the parents' generation, for example, the researchers saw a correlation between early death in men and the presence in their children (and therefore presumably in the parents) of a *nicotine receptor allele* that makes it harder to quit smoking. Many of the men who died young had reached adulthood in the United Kingdom in the 1950s, a time when many British men smoked 20 cigarettes a day. In contrast, the allele's frequency in women and in people from northern California did not vary with age, presumably because fewer in these groups smoked heavily and the allele did not affect their survival. As smoking habits have changed, the pressure to weed out the allele has ceased, and its frequency is unchanged in younger men. Indeed, Pickrell's team detected other shifts. A set of gene variants associated with late-onset menstruation was more common in longer-lived women, suggesting it might help delay death. Also the frequency of the *ApoE4 allele*, which is associated with Alzheimer's disease, drops in older people because carriers died early. Selection during an individual's life span point to the power of large studies to understand the factors that determine survival and reproduction in humans in present-day societies.

Fallen fruit has been a source of nutrition for millions of years. However, fruit that has not been collected after falling to the ground will rot, and reveal its presence by its smell. This may be due to alcohol, which has developed as a consequence of fermentation of naturally present sugar. About 10 million years ago, a common ancestor of humans, chimpanzees, and gorillas evolved a version of the protein that was 40 times more efficient at ethanol metabolism. Without the *alcohol dehydrogenase* (ADH4) mutation the ethanol would quickly build up in the blood stream leading to inebriation and reduction in life-span for health reasons and susceptibility to natural predators. Researchers didn't know how long ago people evolved the more active form of the enzyme. Some scientists suspected it didn't arise until humans started fermenting foods about 9,000 years ago.

Today the conditions of modern life could be driving changes to genes for certain behavioural traits, for example alcoholism (*ADH1B and ALDH2*) autism (*SHANK3* gene), depression (*serotonin transporter gene* (SLC6A4), drug addiction *(D2 subtype of dopamine receptor)*. This has been studied at length in association to substance addiction. The D2 receptor responds to the chemical dopamine which produces rewarding and pleasurable feelings in the brain. Through mice studies, agreeing contemporary research has shown that individuals with a deficiency in this dopamine receptor exhibit not only a preference for and increased consumption of alcohol over their genetically normal peers but also compensated levels of the *cannabinoid receptor type CB1*.

This suggests that both of these genetic factors work together in the regulation of alcohol and cocaine in the brain and in the normal regulation of dopamine. Individuals with this genetic deficiency in the D2 dopamine receptor may be more likely to seek out these recreational pleasure/reward producing substances as they are less receptive to the natural "feel good" effects of dopamine. This naturally occurring deficiency is one of the most studied genetic vulnerabilities to substance abuse across the field. Recent studies show that *gamma aminobutyric acid* (GABA) also plays a role in vulnerability to addiction.

This raises the fascinating possibility that some individuals possessing a deficiency of the D2 dopamine receptor, may as a consequence feel less happy, and therefore when perchance experiencing the longed for pleasure to which they have hitherto been denied, are thereby increasingly prone to relive this sensation by taking certain recreational drugs. This rings a bell regarding vitamins.

Not only is vitamin addiction possible, but it can have very real consequences. Many consumers believe that supplements are harmless, but these substances can cause just as much harm to your body as some of the worst illicit drugs.

Vitamin addiction almost always begins with innocent intentions. The average consumer is highly invested in his or her health, and the general consensus is that taking supplements can improve your wellbeing.

So may it be that some D2 dopamine receptor deficient individuals are simply self-medicating by taking recreational drugs?

Drinking heavily puts people at risk for many adverse health consequences, including alcoholism, liver damage, and various cancers. But some people appear to be at greater risk than others for developing these problems. Why do some people drink more than others? And why do some people who drink develop problems, whereas others do not?

It may therefore be inferred that the evolutionary change that resulted in the expression of the alcohol dehydrogenase 4 gene meant that humans found alcohol in general to be better tolerated and in many instances, desirable. This in turn would have, and has, exerted a selective effect on the genetic constitution of humans.

Research shows that alcohol use and alcohol-related problems are influenced by individual variations in alcohol metabolism, or the way in which alcohol is broken down and eliminated by the body. Alcohol metabolism is controlled by genetic factors, such as variations in the enzymes that break down alcohol; and environmental factors, such as the amount of alcohol an individual consumes and his or her overall nutrition. Differences in alcohol metabolism may put some people at greater risk for alcohol problems, whereas others may be at least somewhat protected from alcohol's harmful effects. The basic process involved in the breakdown of alcohol, will be described.

Alcohol is metabolized by several processes or pathways. The most common of these pathways involves two enzymes – *alcohol dehydrogenase* (ADH) and *aldehyde dehydrogenase* (ALDH). These enzymes help break apart the alcohol molecule, making it possible to eliminate it from the body. First, ADH metabolizes alcohol to *acetaldehyde*, a highly toxic substance and known carcinogen. Then, in a second step, acetaldehyde is further metabolized down to another, less active byproduct called *acetate*, which then is broken down into water and carbon dioxide for easy elimination. If there are bacteria present e.g. *acetobacter* in air, on fruit, on grape presses, the wine can slowly be turned into acetic acid (vinegar).

Some researchers believe that acetaldehyde may be responsible for some of the behavioural and physiological effects previously attributed to alcohol. For example, when acetaldehyde is administered to lab animals, it leads to incoordination, memory impairment, and sleepiness, effects often associated with alcohol.

Genetic differences in these enzymes may help to explain why some ethnic groups have higher or lower rates of alcohol-related problems. For example, one version of the ADH enzyme, called ADH1B*2, is common in people of Chinese, Japanese, and Korean descent but rare in people of European and African descent. Another version of the ADH enzyme, called ADH1B*3, occurs in 15 to 25 percent of African Americans. These enzymes protect against alcoholism by metabolizing alcohol to acetaldehyde very efficiently, leading to elevated acetaldehyde levels that make drinking unpleasant. A recent study found that two variations of the ALDH enzyme, ALDH1A1*2 and ALDH1A1*3, may be associated with alcoholism in African-American people.

Ironically, the very genes that protect some people from alcoholism may magnify their vulnerability to alcohol-related cancers. The International Agency for Research on Cancer asserts that acetaldehyde should be classified as a carcinogen. Acetaldehyde promotes cancer in several ways – for example, by interfering with the copying (i.e., replication) of DNA and by inhibiting a process by which the body repairs damaged DNA. Studies have shown that people who are exposed to large amounts of acetaldehyde are at greater risk for developing certain cancers, such as cancers of the mouth and throat. Although these individuals often are less likely to consume large amounts of alcohol, Seitz and colleagues suggest that when they do drink their risk for developing certain cancers is higher than drinkers who are exposed to less acetaldehyde during alcohol metabolism.

Researchers continue to investigate the reasons why some people drink more than others and why some develop serious health problems because of their drinking. Variations in the way the body breaks down and eliminates alcohol may hold the key to explaining these differences. New information will aid researchers in developing metabolism-based treatments and give treatment professionals better tools for determining who is at risk for developing alcohol-related problems.

Animal and plant breeders are of course well aware of the paths available to vary and improve the species, so it is natural to pose the question, "Can we improve the human breed?" Options may include gene therapy applied to cure of a disease affecting a particular organ, or the entire individual by altering the germ line. Extreme care will be needed, because our knowledge of the desired effect on one gene need not necessarily apply to other functions subserved by this same gene.

Genetics holds particular promise for meeting the challenge of adapting to a warmer world in agriculture. Nature Genetics co-organised a conference with New York University, titled *Plants of the Future*. The focus of the meeting was how to best meet the twin challenges of human population growth and climate change through genetic, genomic or other biological approaches, and to envision what the 'plants of the future' might look like. There is a pressing need for sustainable ways to increase the efficiency and yield of staple crops, and we are excited to see the energy and ingenuity being devoted to this issue. Specifically, using genetics and genomics methods can accelerate the identification or generation of crop plants that can survive and thrive in changing environments.

However one must not forget the serious impact on the human frame of global warming.

Heat stroke is a life-threatening illness commonly found in tropical areas and during hot seasons elsewhere. It is characterized by an elevated core body temperature above 40°C, which induces multi-organ failures (such as circulatory shock, central nervous system dysfunction, acute renal and liver failures, etc.) and sometimes followed with heat cytotoxicity, coagulopathies and systemic inflammatory response syndrome. Ambient temperatures approaching 50°C are likely in the near future in some areas on Earth. Heat stroke is experienced primarily by immunocompromised individuals, such as the very young and the elderly. The inability to properly predict, diagnose and treat long-term sequelae of heat stroke is a serious concern of modern medicine, which reflects our limited understanding of its pathophysiological mechanisms mediating tissue injuries. Recent epidemiological studies of short- and long-term heat stroke outcomes indicate that *multiple organ dysfunctional syndrome* continues to manifest in patients following clinical treatment and increases mortality during the ensuing months and years of recovery. Therefore, application of novel technologies, including radio-telemetric, genomic and proteomic analyses, in heat stroke research is of essence in advancing our knowledge on its pathophysiology. The association between heat stroke and the *calsequestrin 1* (CASQ1) gene was reported in a recent study, which represented a significant step forward. Future research into possible genetic manipulation enabling better tolerance of exposure to higher ambient temperatures may well assume greater importance in the not too distant future. The future of survival on our planet may well rest with genetic research and genetic selection, both by Nature and manipulation.

Chapter 15

Zoroaster (? 1500-1000 BC)

Zoroaster

Zoroaster, also known as Zarathustra, Zarathushtra Spitama or Asho Zarathushtra, was an ancient Iranian prophet, religious reformer and spiritual leader who founded what is now known as *Zoroastrianism*. The Oxford Dictionary of Philosophy considers Zoroaster to have been the first philosopher. He was the founder of the national religion of the Iranian people. *Zoroaster* is the corrupt Greek form of the Iranian *Zarathustra*.

Below is a copy of an article, as explained. I quote:

> This article is also featured in the FEZANA journal of Summer 2001. Journal of Zoroasterianism and Science. Reproduction of the article has been done with the permission of the author and that of the guest editor of the Summer 2001 issue of the FEZANA journal, Mrs. Dina McIntyre. Religion and Science Theology.

Author Dr Lovji D. Cama. Here is the text:

Albert Einstein once said: "If something is in me, which can be called religious then it is the unbounded admiration for the structure of the world so far as our science can reveal it."

While some of his other quotes suggest that he did not believe in a personal God, Einstein considered himself to be religious. The "structure of the world" that amazed Einstein is analogous and possibly synonymous to what we as Zoroastrians call Asha.

A personal God, who rules at will and the workings of whose mind cannot be comprehended or questioned is a concept that is rather difficult for a scientist to accept. What about the Zoroastrian concept of Ahura Mazda? Here is a concept of God whose main attributes are absolute wisdom and absolute goodness, a God who is incapable of evil and in that sense not absolutely powerful. This is a God who is completely predictable. Ahura Mazda, who is Vohu Mano (the Good Mind) and Asha (Righteousness and the Divine Law), has a perfect plan for the universe that he has created. The nature of this perfect plan is subtle and intricate but it is not capricious. Its intricacy arises from the infinite nature of its Creator but it can be understood and unravelled, though perhaps not completely. The understanding comes from the use of the Good Mind that has been granted to mankind.

This endeavour of understanding Asha is science when applied to the physical world and is religion in its truest sense when applied to the moral world. When we look at the universe in this way the distinction between science and religion becomes blurred and the conflict disappears.

Where does the conflict between religion and science come from? In the absence of knowledge about the physical world, religion creates myths that temporarily may satisfy humanity's need for understanding the mystery of the physical world. Over time, these myths do not stand up to critical examination via a scientific approach. When religion insists on the validity of these myths, there is conflict. Such is the conflict between the theory of evolution and the Biblical myth of special creation. In fact, at face value, most of the stories of creation as described by various religions are in conflict with the theory of evolution, including our story of creation in the Bundahishn.

Does evolution then conflict with the teachings of Zoroastrianism?

In my opinion, no. It is possible to take these myths, not literally, but as attempts to explain the nature and origin of man in God's plan, in which the case the conflict goes away. Evolution in my mind is the Law of Asha unfolding to create mankind, this is the real story of creation. The laws of nature which are part of the Divine Law of Asha, have created a sentient being that is aware not only of itself, but also the creation around it. A being willing to attempt to understand the Creator, and if the Law of Asha is the Plan of God for the universe, then evolution is the only way this could have happened.

The pursuit of science is the pursuit of understanding the mystery of the physical world. Two of the greatest mysteries are the origin of the universe and the nature of life.

According to the big Bang theory, our universe started from a point of inconceivably immense energy and following the predictable laws of physics gave rise to all the stars and galaxies as we know them today. One cannot but be amazed at the order that is present in the universe. Biologists and physicists alike have found that our universe is very finely tuned. The slightest of deviations of the laws of physics would have prevented the events leading to our present state and would have produced a very different universe in which humans would not have evolved. If our universe is a product of intelligent design, then this order that one can see and measure is the order introduced into the universe by the designer, for a Zoroastrian, this is a manifestation of Asha.

In the last hundred years man has made great strides in understanding the nature of life, to the extent that the entire genomes of quite a few species, including our own, have been mapped. Though the blueprint for life appears to be laid out in a very simple fashion, we have not been able to figure out all the mechanisms by which even a simple bacterial cell functions. That is not to say that we do not know many of the biochemical steps. Even our limited knowledge has been useful, for example, in defending ourselves with antibiotics against bacteria. However, the more we learn, the more there appear to be phenomena left to discover. The complex systems of chemical reactions that are a living cell require an order that we can only begin to decipher. In the absence of this order, there is no life and as we have said, for a Zoroastrian, this order is a manifestation of Asha.

Ecology is a relatively new field of science. Its genesis is the belated realization that man in his progress has disregarded the effects of his endeavours on the rest of nature. The success of this species has taken a great toll on other species whose habitats man has destroyed. Man has come to the realization that the power that science has given to man over nature must be used wisely, i.e. with the Good Mind to establish Kshathra Variya (a just dominion, in which all species can thrive). Ecology is our attempt to do that. Zoroastrianism teaches us a great respect, and even a reverence for nature. This is demonstrated by our rituals and our prayers dedicated to various aspects of nature. If the world would have the same reverence for nature that Zoroastrians are supposed to show, ecology would always be a part of the decision making process.

Science, the methodical pursuit of trying to understand the universe as perceived by our senses, and religion, the attempt to understand our Creator and the universe through our spirituality, are not mutually exclusive. Especially for Zoroastrians, science and religion answer questions at different ends of the spectrum of Asha, The Divine Order. Often, when their spheres overlap, they provide each other with important insights and

> *thought-provoking questions. For an example of this, we can return to the Big Bang theory: where did the initial energy come from? Is it God who put it there? Or even more stirring, is that energy a manifestation of God? Finally, as Zoroastrians, we are expected to use our Good Mind to reconcile our ancient myths with our new scientific findings.*

To me, the author of this book, I feel it sums up in a succinct and ingenious manner, the bridge between science and religion. Although I am not personally a Zoroastrian I hope that you, the reader, will perceive, through the dense mist, the outlines of the connection. And accordingly, a way ahead.

Bibliography

Charles Darwin's The Life of Erasmus Darwin, (2003) Edited by Desmond King-Hele, (Cambridge University Press).

The Darwins of Shrewsbury, (2009) Andrew Pattison, (The History Press).

On the Origin of Species, (2004) Charles Darwin, (Collectors Library).

Anna Seward's Life of Erasmus Darwin, (2010) Philip Wilson, Elizabeth Dolan, and Malcolm Dick (Brewin Books).

Erasmus Darwin: A Life of Unequalled Achievement, (1999) Desmond King-Hele, (Giles de la Mare).

The Voyage of the Beagle, (2015) James Taylor, (Conway, Bloomsbury).

The Descent of Man; and the Selection in Relation to Sex, (2009) (Digireads.com).

The Descent of Darwin, (1982) Brian Leith, (Collins).

The Voyage of the Beagle, (1997) Charles Darwin, (Wordsworth Classics).

Epigenetic Inheritance and Evolution: The Lamarckian Dimension, (1995) Eva Jablonka and Marion J. Lamb, (Oxford University Press).

Erasmus Darwin: Sex, Science and Serendipity, (2012) Patricia Fara, (Oxford University Press).

Cosmologia: A Sequence of Epic Poems, (2004) Erasmus Darwin, (Stuart Harris, Sheffield).

The Evolutionists: The Struggle for Darwin's Soul, (2001) Richard Morris, (Henry Holt and Company, LLC).

Richard Owen: Biology without Darwin, (2009) Nicolaas Rupke, (University of Chicago Press).

The Stress of Life, (1976) Hans Selye, (McGraw-Hill Book Company).

Index

By the Same Author

The Life and Times of William Withering: His Work, His Legacy, (2004).
The Fall and Rise of Aspirin: The Wonder Drug, (2007).
Steroids in Society: Use and Abuse, (2013).

All available to purchase at: www.brewinbooks.com